LUCY SIEGLE is a journalist, broadcaster and expert on climate and nature issues. She was the first dedicated eco columnist to be appointed to a national newspaper in the UK; her Ethical Living column ran in the *Observer* for over a decade. On TV she is known for reporting on every aspect of the planet for BBC1's *The One Show* and giving clear actionable advice on living with a lighter footprint. Her previous books include *Turning the Tide on Plastic* and *To Die For*. The latter was turned into hit Netflix documentary *The True Cost* (on which she is co-executive producer with Livia Firth). She is a trustee for the environmental NGO Surfers Against Sewage (SAS) and co-hosts a podcast, *So Hot Right Now*, about storytelling the climate crisis. She also works on climate campaigning with singer-songwriter Ellie Goulding, as well as other high profile environmentalists.

You can follow Lucy on Twitter @lucysiegle
and on Instagram @theseagull

T0007604

ALSO BY LUCY SIEGLE

Turning the Tide on Plastic
To Die For

BE THE
ULTIMATE
FRIEND OF THE
EARTH

100 QUESTIONS TO BOOST YOUR CLIMATE AND NATURE IQ

LUCY SIEGLE

Michael O'Mara Books Limited

First published in Great Britain in 2022
by Michael O'Mara Books Limited
9 Lion Yard
Tremadoc Road
London SW4 7NQ

A CIP catalogue record for this book is available from the British Library.

Papers used by Michael O'Mara Books Limited are natural, recyclable products made from wood grown in sustainable forests. The manufacturing processes conform to the environmental regulations of the country of origin.

ISBN: 978-1-78929-393-7 in paperback print format
ISBN: 978-1-78929-396-8 in ebook format

1 2 3 4 5 6 7 8 9 10

www.mombooks.com

Cover design by Claire Cater
Designed and typeset by Claire Cater
Illustrations by Peter Liddiard

Printed and bound by CPI Group (UK) Ltd, Croydon, CR0 4YY

MIX
Paper from
responsible sources
FSC® C171272

CONTENTS

INTRODUCTION

Welcome to this mission to become the ultimate friend of the Earth. I'm already impressed by you. By picking up this book it proves to me that you are keen to do everything you can to look the planet straight in the eye and admit that we haven't done brilliantly thus far, but this is where things change for the better. When I say, 'we haven't done brilliantly thus far', of course that's an understatement. It's become common knowledge that we humans have put ourselves on a collision course with our own planet. Without a complete rethink of the way we live and behave and a switch to different systems that replenish the resources on Earth rather than constantly exploiting them, vast parts of the planet are likely to become uninhabitable to us. I'd call that an emergency, wouldn't you?

Over the last few years, we've started to describe the emergency we face with greater accuracy and honesty. Although that is extremely scary, and personally makes me feel anxious, I'd much rather we talked about it than pretended it's going to go away. It's not going away until we find ways of solving it.

Throughout this book we'll talk about a crisis or emergency in climate and in nature. You've probably heard these talked about separately, but here's why I think they should go together: they are inextricably linked. We know that the climate is changing fast. Greenhouse gases (of which carbon dioxide is the most significant, and perhaps the most famous) have been building up in the atmosphere since the Industrial Revolution. Humans started burning large quantities of fossil fuels to power their progress. We are still doing much the same today! The rate of burning has intensified, causing gases to accumulate in the atmosphere. There they act like a blanket, trapping heat around the Earth (the so-called 'greenhouse effect'). This is causing the surface temperature of the Earth to increase. In turn, this changes everything for the worse, and that's why we call it a climate crisis. More heat means melting ice, different weather and more acidic oceans. Extreme weather events such as hurricanes become more common. Sea levels rise, contributing to storm surges that cause epic flooding and chaos. On the flip side, droughts and wildfires become more common. All this threatens various systems that help to sustain our lives. For example, the climate crisis impacts on agriculture, making it harder to produce the food we need. While most of us have become used to warnings about the danger from the build-up of greenhouse gases, what we're often less familiar with is the way that these greenhouse gases threaten biodiversity – a word that describes the biological variety of living organisms on Earth (I think of it as a formal name for 'nature', and that's the term I'll mainly use in this book).

In a nutshell, climate change makes all the functions of nature work less well. This is a serious problem, because nature underpins everything – the water cycle, for example, gives us fresh water to drink, the nitrogen cycle gives us fertile soils in which to grow our crops. A climate in crisis impacts on these systems, creating a nature crisis too. Then, to make matters worse, we panic. We start to convert more land such as forests to grow more crops, to make up for the fact that our food isn't growing so well. Big mistake. Not only does this destroy the homes of other people – especially indigenous communities who traditionally live alongside nature successfully – but it destroys the habitats of other creatures. Meanwhile those trees you cut down (and I don't mean 'you' specifically; don't worry, I'm not shouting at you!) would otherwise do an incredible job in sucking up carbon dioxide (the main greenhouse gas emitted) and using that carbon to create new growth for more trees. By cutting down so many trees, we're removing the most important carbon sink we have. So you'll see how a crisis of climate rolls into a crisis of nature, and how they become one huge emergency. If we're honestly trying to change our fortunes, we have to face up to the fact that they cannot be separated. That's despite the fact that it might seem like double the workload to solve two huge problems. But I prefer to look at it a different way: to realize that we are the first humans on Earth to know and acknowledge this. That is very powerful because, yes, we face a huge, potentially catastrophic emergency, but we also know more about how to solve it and have more tools at our disposal to do so than at any other time in human history. However, most of the world's expert scientists agree that, to be effective, we need to address climate and nature together. So let's pledge to put them together, at all times.

Knowing how to respond to this dual emergency seems really complicated. My theory is that this is because we have simply forgotten how to be a great friend to the Earth – in which case we need to relearn or give ourselves the chance to do better. But who do we learn from? Well, unfortunately, often from the wrong people. You may have noticed, for example, that some of our political leaders do not seem to have a very good relationship with our planet. They are stuck in old patterns of behaviour in which, instead of respecting the Earth's boundaries (as we'll discover, these are pretty important), they just want to remove more and more of the planet's resources and at an ever-faster speed. They mistakenly think this will make everything better, whereas in fact it will actually make things worse.

Meanwhile, real change can seem out of reach, out of our price range and even out of this world. If you're constantly hearing that the only way to be a great friend to the Earth is to buy an electric sports car, and you can't afford an electric sports car or don't know how to drive, you're going to feel excluded from being a great friend. Similarly, if you feel that you have to understand all the technical details of the Paris Climate Agreement but don't excel at details or find the minutiae of such stuff particularly dull, that's pretty excluding too. This is a great time to shift your focus to defending, protecting, enjoying and understanding Planet Earth. Because, while I know there's a lot of doom and gloom around, it's also true that we currently know more and are more confident about our relationship with the living planet than at any other time in human history. Thanks to an accumulation of scientific understanding and the research and dedication of visionaries from past generations, we are now at the point where we can say with certainty that the ocean drives climate, and is a critical part of the carbon cycle; that the great forests create rain that waters crops hundreds of miles away; and that millions of living organisms interact to stabilize conditions on Earth. We know more about the harm we are doing, but we also know more about how this incredible planet works. We are, after all, the first humans to have grown up knowing what the Earth looks like from above. If we educate ourselves and allow ourselves to absorb the science, we can only be guided by it.

This book, therefore, is about reasserting and rethinking ways to boost our friendship credentials. My feeling is that if we regard the Earth as a great friend/amigo/pal throughout our lives, it will seem natural to make decisions that are Earth-friendly rather than the ones we all typically make at the moment, which tend to poison it, steal its stuff and treat it as a massive planetary doormat. Throughout my adult life I've been trying to think of ways of making sense of the countless incredibly big themes and worries concerning the environment. But don't feel sorry for me, because I feel lucky. I get to think about these issues 24/7, while most of my friends have to cram in eco-anxiety while they're doing all the usual stuff you have to do in life – working, picking their kids' toys off the floor, ordering stuff online, keeping the fridge stocked, trying to stop the dog eating their tennis shoes – you know, ordinary, everyday stuff.

It was through thinking about my friends that I came up with the idea of this book. I thought about how we interact, support, jostle each other, keep one another in check, balance, even out and regulate each other, and I came to realize that all we have to do in order to shift our mindset from a negative one – where our action on climate and nature is stunted by fear, anxiety and confusion – to a positive one – in which we do everything we possibly can to look out for the Earth – is to treat the planet as our best friend. So, for the duration of this book, keep at the front of your mind the idea that you are going to up your game from being an indifferent, or perhaps even exploitative, citizen of the Earth (whichever best sums up your current relationship) to becoming a 100-per-cent committed ally who will be there for your number-one amigo at every twist and turn.

I thought it would be fun to include a quiz element in the book. There are two reasons for this. First, everyone loves a quiz. OK, everyone *I* know, at least, loves a quiz. But second, there is a lot of information involved in this subject (we are talking about the entire planet, so that stands to reason) and adding an element of competition (with yourself or other people) significantly helps to make that information stick in your brain.

This brings me to another point. Would you consider someone a good friend if you were totally clueless about them? No, I thought not. But when it comes to the Earth – which, seriously, is the most important collaboration of our existence – we are often pretty clueless. This isn't always our fault. Parents and kids (and even teachers) often complain to me that they don't get to study enough about climate and nature at school. If we are to be the ultimate friend of the Earth, we cannot continue to ignore, minimize or misrepresent it. We must understand as much as possible about what makes the Earth tick and flourish, and, conversely, what harms and destroys it. So this book is full of information about our planet, how it works and how connected our fortunes are with all other living things.

This is not a hefty academic primer on earth systems and ecology; I hope, rather, it will be fun and eye-opening. It's only when our eyes are opened and we're engaged that we become really creative and resourceful. And it's when we humans are firmly in that frame of mind, and poised for action, that we become very useful in a crisis. So make no mistake, despite the purposely jolly tone of this book, we are in a very deep crisis affecting our beautiful planet and threatening our futures. But armed with knowledge about how to reverse that and our combined fortunes, and how to build a really beautiful and worthwhile future, all is not lost. Are you with me? I hope so!

HOW TO USE THIS BOOK

This book divides neatly into ten chapters, each focusing on a particular subject. It probably makes sense to read them in order, but feel free to go where the mood takes you. In each you'll find ten quiz questions. They take inspiration from the great biomes (the name given to a big collection of animal and plant species that occupy a major habitat and include desert, tropical rainforest, temperate grassland and boreal forests) and spectacular and intriguing flora and fauna, from tiny insects to huge personalities including Earth activists, high-profile tree-huggers and leaders of global movements. Keep a piece of paper and a pencil handy, especially if you are going head-to-head with friends and family. That way you can tot up your score as you go along.

Each chapter begins with an introduction and some hopefully useful information, so that you can become an armchair expert on everything from how brilliant oceans are at being carbon sinks to the circular economy that's going to make the stuff we use less impactful.

Then we jump into our questions. Be warned, some are challenging and there are decoy answers in there too. It's designed to stretch your grey matter and test you thoroughly. After all, the prize is true friendship with the actual biosphere (the zone of all life on Earth); we can't give that away too easily.

Immediately after the questions you'll find the answers. They aren't tucked away at the back of the book here. Rather, I want you to read them while the questions are still pinging around in your brain. The answers should give a lot of extra insight into the chosen topic, so you can really start to put together a picture of how our incredible planet works and what you can do to keep it wealthy and healthy. There's no shame here if the first time through you don't do as well as you hoped. I hope you'll return and quiz yourself (and admire how much you now know) many times, so don't feel you have to do everything in one sitting. But I really hope you get to compete against friends or family. Research shows that our positive environmental behaviours increase when we're in competition with others – it's a case of human nature helping actual nature.

HOW TO GET INTO THE ULTIMATE FRIEND ZONE

There are a hundred diverse questions in this book to test your existing knowledge (and love) for the planet. After the final round, I'll ask you to tot up your overall score. Based on that score, you will fit into one of five different zones. This is what your overall score gets you:

1–20: You've reached the first rung on the ladder, but you are definitely on the ladder! Before long, you will have developed a whole Earth-friendly mindset.

21–40: You've made an assured start and are on solid ground. You've got some Earth-defending skills but now you need to nurture them.

41–60: You've reached our equivalent Everest base camp. A great achievement in and of itself, but now you need to aim for the summit!

61–80: You're an unashamed Earth science nerd, tuned into the workings and needs of a dear friend. You are a powerful ally and, with one more push forward, you'll be up there with the greats.

81–100: You've done it. You've proved yourself worthy of being crowned the ultimate friend of the Earth. This is your green coronation!

Even if you are a bit disappointed with your score the first time through, by the time you get to the end of Chapter 10, you will have accumulated knowledge from a hundred questions and ten chapters stuffed full of ideas. Whatever your 'official' score, that's the point at which you can really declare yourself the ultimate friend of the Earth. I just have one unofficial question before we begin: *are you ready?*

1

PLANET HYPE!

In the context of *The Simpsons*, Planet Hype is an 'over-hyped, overly trendy movie-themed restaurant located at the Springfield Squidport'. In this book, it's an instruction – hype the planet! Why? Because Earth needs a cheerleader, and I nominate … you!

In fact, it's surprising we don't hype or cheerlead for the Earth all day long. Because, to borrow from another cultural juggernaut, Dorothy in the *Wizard of Oz*, there's no place like home. The Earth is our home, and we know no other (no matter what anybody tells you about their interactions with extra-terrestrials).

That makes us incredibly lucky. The conditions on our planet are just right for life to flourish, and, until recently, pretty ideal for our species, *Homo sapiens* (this is a bit of a spoiler, but we'll learn in the next chapter that the situation is changing fast). Most of the time we think about our immediate environs. Today, my curiosity about the planet will extend only as far as two blocks away when I take my small terriers – Daphne and Arthur – to the park. But if we want to maintain a decent relationship with the Earth then we'll need to take a wider view as often as possible.

Often we wait for big events to prompt us into doing this, or we live vicariously, waiting for other adventurous humans to report back. An example of this would be waiting for astronauts who have secured a bird's-eye view of Earth to tell us how wonderful it is. Well, there's nothing wrong with that. Such a perspective is well worth listening to. But we shouldn't just rely on eyewitness accounts from very lucky or exceptional humans. Perhaps you too hope to become an astronaut; but just in case that doesn't work out, why not try to discover something spectacular about the Earth every day, week or month (pick the most realistic timespan). Seek out some information under your own steam. The more you know, the better you'll be able to hype the planet and the better friend you will be (plus you'll do fantastically well in our quiz questions).

Did you know?

The Earth has never been perfectly round. I'm not saying it's flat, but according to NASA it bulges around the equator by an extra 0.3 per cent. This is a result of the fact that it rotates about its axis.

Anyway, let's take a minute to think about how incredible our home biosphere (an umbrella term for all life) is. An estate agent would describe it as 'highly desirable'. There is a great distance between our planet and the sun – not too hot, not too cold (most of life flourishes in temperate climates), and we have water in liquid form. If we did not have water in liquid form, we would soon know about it because it is the single most important factor for supporting life. The planet also has some other useful attributes, including being just big enough to have sufficient gravity to hold on to our atmosphere. The atmosphere (you might know it better as 'air') surrounds the Earth like a thin skin. If we lived on a smaller planet, such as Pluto, and didn't have as much gravity, we wouldn't be able to hold on to the atmosphere around our planet and it would drift off into space. Fortunately, our planet is not too big, and not too small.

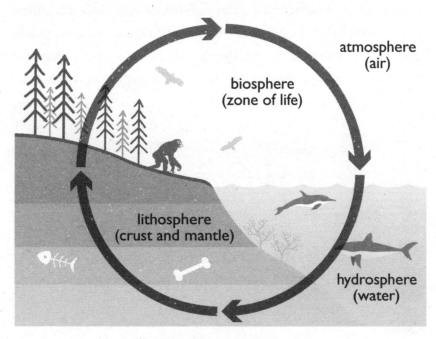

The four spheres of the Earth.

Now, obviously, we don't go around all day thinking about how lucky we are to be just the right size and distance from the sun to maintain our existence. But hopefully we do notice other stuff. We know that the biosphere captures and holds some pretty amazing biomes with individual characteristics. I like this description from the *National Geographic*: 'The biosphere extends from the deepest root systems of trees to the dark environment of ocean trenches, to lush rain forests and high mountaintops.'[1] I hope by the end of this book, you'll have your own favourite descriptions of geeky terms such as biosphere (if you don't already). Enthusiasm for the Earth is definitely part of this mission. But in order to become a better ally to the planet, we need to dial up these connections and acknowledge a couple of major points: first, the Earth is not just a lump of rock for us to stamp all over. We can only push it so far and there's plenty of research that suggests the Earth has evolved to work hard to stabilize these great conditions to sustain life. There are all manner of incredibly complex mechanisms and safeguards that the Earth deploys to keep things in balance: forests, oceans and other elements, which help to maintain a fairly constant surface temperature and are near optimal for life.

Did you know?

Unexpectedly, the driest place on Earth – the Atacama Desert (northern Chile) – is slap-bang next to the biggest body of water on Earth, the Pacific Ocean.

The Earth is not here as a sort of grab-and-go buffet, a smorgasbord of resources for us to help ourselves to. But, as we'll also investigate later on, we've developed some outrageous habits and outlooks that are really testing our relationship with it. Instead of taking and scheming to take more, it's time to turn this approach on its head and pay the Earth some compliments. Go on, have a go right now. What is brilliant about this planet? Which biosphere gets your vote? Say thank you for something the Earth has given you. Do you have a favourite biome? Is there one ecosystem that you're particularly interested in (even obsessed by)? Some humans make it their business to relocate to the ecosystems, habitats or terrains they love. Some of us remain quietly adoring from afar. In the future our travel might be restricted because of the need to reduce greenhouse gas emissions. There is already a flourishing No Fly movement. On the flip side, millions of people will need to move, forced to flee the effects of ecological change and the climate crisis. Some ecosystems will become difficult to live in and there is only so much adaptation that species can do. But that's for later in this book; for now we're focusing on what makes Earth special.

The point is that we are privileged to call this planet home because it's super-impressive. I'm not really in the Jeff Bezos camp of blasting off to another planet, because I'd rather spend the money and time correcting this one.

OK, let's move on to the first quiz to find out how clued up you are about Earth's brilliance, brains and beauty. If you score 3/10 or under, I suggest you ask Jeff for a lift to space.

TIME TO SHOW OFF YOUR TERRESTRIAL TRIVIA
YOUR ESSENTIAL TEN-QUESTION QUIZ

1. It wasn't until 1972 that humans got a true bird's-eye view of Planet Earth when the *Apollo 17* crew were able to get a photograph from the space station. The image prompted the creation of the first ever Earth Day and is synonymous with ecological action, but what is that iconic photograph known as?

A. The Blue Marble
B. The Living Planet
C. The Blue Planet
D. The First Eden

2. In this chapter we've discussed how your friend Earth works hard to provide the right conditions for life. What is the experts' nickname for the very habitable (highly desirable) area around the sun in which the Earth orbits?

A. Waterworld
B. Daisyworld
C. Arcadia zone
D. Goldilocks zone

3. We could hardly talk about how amazing the biosphere is without referencing the stretch of thirteen major islands containing five different habitat zones that has become a mecca for anyone interested in natural history. But what were the Galapagos Islands called before?

    A. Los Pollos Hermanos
    B. Los Abundantos
    C. Las Encantadas
    D. Las Tortugas Gigantes

4. We've blanked out the ecosystem in this description, can you put it back in? 'Despite its clarity and simplicity, the _____ wears at the same time, paradoxically, a veil of mystery. Motionless and silent, it evokes in us an elusive hint of something unknown, unknowable, about to be revealed.' What ecosystem is being described here?

    A. Desert
    B. Forest
    C. Ocean
    D. Grassland

5. The cryosphere is the term used to describe the incredible amounts of solid water that the Earth holds in different forms – glaciers, ice caps, ice sheets, snow, permafrost, and river and lake ice. There is some debate as to whether it should be described as a biome in its own right![2] Which part of the cryosphere does the following short extract describe?: 'With an average elevation of over 4,000 m, it is the highest and the largest highland in the world and exerts a great influence on regional and global climate … Its surroundings contain a large number of glaciers which are at the headwaters of many prominent rivers.'[3]

    A. The Third Pole
    B. The North Pole
    C. The South Pole
    D. The Arctic Eight

6.

Let's stay with the cold. What is the real name of the glacier that is sometimes referred to as the Doomsday glacier due to the fast rate at which it is melting and its potential contribution to an increase in global sea levels?

A. Larsen B
B. Thwaites
C. Majestic
D. Zachariæ Isstrøm

7.

Scientists estimate that, from top to bottom, the biosphere is 20 kilometres in height, around 12 miles. But what do they use to determine the top of the biosphere?

A. The point at which a satellite goes into orbit
B. The point reached by the highest-known flying bird
C. The point at which aeroplane flight-control systems are no longer effective
D. Super Low Earth Orbit is used as the measure

8.

The Amazon rainforest is the largest tropical rainforest containing an incredible 40,000 plant species and 2.5 million different insects. It also provides an extraordinary portion of the everyday stuff we rely on, from chocolate to our medicines and moisturizers, and stuff we hardly even think about, like food colouring. But do you know how many countries the Amazon rainforest spreads across? (There's a bonus point if you can name them all!)

A. Three
B. Four
C. Six
D. Nine

9. Over the years enterprising, green-minded humans (ecopreneurs?) have tried to make their own mini-versions of the planet. In the early 1990s a giant terrarium described as a cross between a greenhouse and the Taj Mahal became home to eight people. But what is the name and location of this mini-Earth?

A. The Eden Project, Cornwall
B. World 2, China
C. Biosphere 2, Arizona
D. Expo World Palace, Dubai

10. 'People protect what they love, they love what they understand and they understand what they are taught.' Who said these wise words (that sum up the point of this chapter so well)?

A. Mahatma Gandhi
B. Wangari Maathai
C. Jacques-Yves Cousteau
D. Nirmal Purja

. . . .

ANSWERS

1. A: the iconic image of the Earth was taken on 7 December 1972 and is called the Blue Marble. The original caption from NASA is below:

   'View of the Earth as seen by the Apollo 17 crew traveling toward the moon. This translunar coast photograph extends from the Mediterranean Sea area to the Antarctica south polar ice cap. This is the first time the Apollo trajectory made it possible to photograph the south polar ice cap. Note the heavy cloud cover in the Southern Hemisphere. Almost the entire coastline of Africa is clearly visible. The Arabian Peninsula can be seen at the north-eastern edge of Africa. The large island off the coast of Africa is the Malagasy Republic. The Asian mainland is on the horizon toward the northeast'.[4]

All of the other answers are the titles of TV series from the great naturalist Sir David Attenborough.

2. D: Goldilocks zone, whose name reflects the 'just right' conditions on Earth for life. It makes sense when you recall the story: the porridge the interloper Goldilocks ate was not too big, not too small. The bed was similarly the perfect size, etc. (You might need to revisit if it's been a while since you read Goldilocks. Clue: there are three bears!)

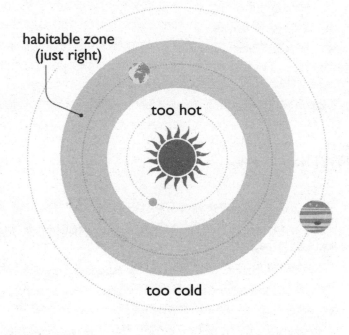

habitable zone
(just right)

too hot

too cold

If you answered B, I can understand what you were thinking. The great chemist James Lovelock (at the time of writing, James is 102 and I can confirm was a very fun lunch companion when I met him at Chesil Beach in Dorset a few years ago) formulated the Gaia hypothesis or theory in 1970 and it was developed by the microbiologist Lynn Margulis. The Gaia theory holds that the Earth is a single, self-regulating system rather than an inert lump of rock. Daisyworld was the rather elegant model used to demonstrate the theory, featuring black and white daisies. White daisies reflect light and heat, black daisies absorb it. But there's only a certain mid-temperature range in which the daisies thrive (reproduce). Too hot or too cold – i.e. out of the temperature range – and they are unable to reproduce and eventually die off.

The Gaia hypothesis was arguably given short shrift by the mainstream scientific community, which was wedded to the notion that life adapts to the environment, not the other way around. This was generally accepted as the case until recently, when a rise in global temperature from climate change triggered some of the complex feedback loops that Lovelock had warned of. Feedback loops are similar to a cyclical chain that repeats again and again. You can think of them as rows of dominoes stretching to infinity. The first domino is pushed forward by something you know about – the rise in global temperature – but this then creates its own reaction that pushes over another and then another. You get loads of ever-increasing reactions that are almost impossible to stop. As Lovelock puts it, 'We have seen just how much life – especially human life – can affect the environment … suffocation by greenhouse gases and the clearance of the rainforests have caused changes on a scale not seen in millions of years.'

3. The answer is C, Las Encantadas, 'the enchanted ones'.[5] If you answered A, you don't speak Spanish and/or were a fan of the TV series *Breaking Bad*. The answer D means 'giant tortoise', and refers to one of the most famous inhabitants of the

Galapagos Islands. Anyway, back to the correct answer. It seems this wasn't meant in a complimentary 'ooh how enchanting' way. 'Man and wolf alike disown them … The chief sound of life here is a hiss', wrote Herman Melville, author of *Moby Dick*, of the islands – quite the Trip Advisor review. The nature apparently was a bit too wild for his liking. But Charles Darwin arrived in 1835 and his resulting book *The Origin of Species* means the rest is history – *natural history* (BOOM!).

However, that's not the end of the story. Nowadays the Galapagos Islands, the cradle of evolutionary science, is frequented by scientists looking for answers about climate change and ecosystem change as increased temperature and rainfall (both consequences of climate change) hit. Will species here adapt or disappear?

4. A: desert, and you get an extra point if you can identify the origin of the quote, because you must be very well read indeed. It's from *Desert Solitaire: A Season in the Wilderness*, an autobiographical work by American writer Edward Abbey, originally published in 1968.

    While life-sustaining provisions and services such as making rainfall are described technically as 'ecosystem services', the atmosphere you get from an ecosystem is known as 'cultural ecosystem services',[6] a phrase that definitely takes the romance out of things. I think Abbey has helped to put the romance back in. While we're on the subject of deserts, did you know they are phenomenal carbon capture sinks and contain a lot more life than you might think? Never write off an ecosystem because it looks a bit dry and dusty; they all have wondrous qualities!

5. This is a description of A, the Third Pole, encompassing the Hindu Kush Himalayas mountain range and the Tibetan Plateau. This part of the cryosphere holds the largest reserve of freshwater outside the polar regions. I've included this question because I've been on a mission to understand more

about the Third Pole. Why? Because when we talk about the climate crisis we tend to focus on the Arctic and the Antarctic and species such as polar bears. But global heating and black carbon pollution from diesel engines and the burning of wood and coal are causing the glaciers at the Third Pole to retreat at a pace described as astonishing, changing the whole water system. Meanwhile, the Third Pole 'feeds' 1.4 billion people through the Indus, Ganges, Brahmaputra, Yangtze and Yellow rivers.

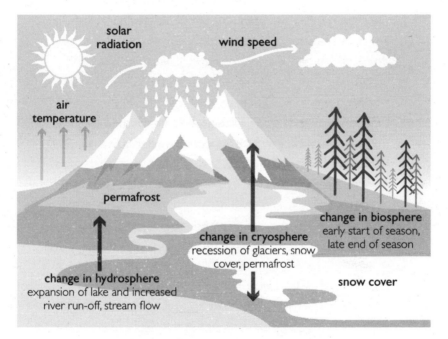

How the biosphere feels the heat.

6. B: Thwaites, another Antarctic glacier that scientists are watching with alarm. You may have opted for Larsen B. Twenty years ago, Larsen B, weighing an estimated 500 billion tonnes, crashed into the sea. That was thought of as critical, but as environmental writer John Vidal puts it, compared

to Thwaites, Larsen B is 'like an icicle'.[7] Thwaites is about a hundred times larger and contains enough water on its own to raise sea levels worldwide by more than half a metre. Satellite studies show it is melting far faster than it did in the 1990s. No wonder it's nicknamed the Doomsday glacier.

7. The answer is B. Although scientists can't be exact about the outer boundary of the Earth's biosphere, their estimates are based on the height flown by the highest-flying bird. That bird is the Rüppell's vulture and it has been recorded flying at 11,300 metres. It makes sense to use the height achieved by a living creature given that the definition of the biosphere is the region where life is sustained. The biosphere's estimated depth meanwhile is based on the depth of the deepest fish, known to live as far down as 8,300 metres in the Puerto Rico Trench.

8. D: congratulations if you answered correctly. For the bonus point you need all of the following: Brazil, Ecuador, Venezuela, Suriname, Peru, Colombia, Bolivia, Guyana and French Guiana.[8]

9. C: Biosphere 2, in Arizona. Number two because obviously the real Earth is number one. This really is a fascinating place. A mini-version of the biosphere, 15 trillion times smaller, it contains different biomes: rainforest, ocean with a coral reef, savannah, desert, mangrove swamp and agricultural fields. Eight (lucky?) people were sealed in this artificial world for two years and the project got off to a rocky start. In the early days, Biosphere 2 attracted the type of scrutiny that reminds me of the attention received by early iterations of reality TV experiments such as *Big Brother*. There were rumours that oxygen was pumped in and question marks over how genuine the experiment was. However, the facility has gone on to provide important insight and data into how warming oceans are killing corals, and to simulate climate impacts on

a half-acre rainforest, giving us vital information on how the planet will fare in a warming world.

10. C: Jacques-Yves Cousteau, a naval captain who turned his hand to telling the stories of the ocean from the helm of *Calypso*, his boat. In fact, it's fair to say that he revolutionized ocean conservation and underwater filming through new techniques that captured the ocean in a way never before seen, brought to TV in a very popular series, *The Undersea World of Jacques Cousteau*. This aired a little before my time, but I'm fascinated by how we can tell stories about the biosphere that show why we are custodians, not owners, of natural resources on our planet.

How did you score in our first round? Don't worry if it was a complete disaster. We have ninety more questions for you to make up the difference. Hopefully you'll take some clues from the first round of questions and answers as you begin to think on behalf of the Earth and employ some Earth logic. If you did well (7 or over) you can put yourself down as a high achiever.

With the words of Jacques Cousteau ringing in our ears, let's push on to Chapter 2.

DON'T BE MEAN
TO THE HOLOCENE

So we've established that this planet is very important to us. Now we must move on to the awkward question of why we are so relentlessly mean to it. A few years ago, I was invited to Stockholm in Sweden to the Stockholm Resilience Centre, which I thought was an intriguing name for a centre (or indeed anything). I was invited to a big meeting chaired by Johan Rockström, who is a bit of a rock star when it comes to sustainability science (the science of keeping the planet in tip-top health, or at least not destroying it under our stewardship). Johan stars in the documentary *The 11th Hour*, narrated by the eco-activist and actor Leonardo DiCaprio (who has spoken out on the climate and nature crisis for almost a decade).

Johan explained the team's work on drawing up planetary boundaries. It is complex – as you would imagine, given the size and variety of the Earth's ecosystems and all the reactions and interactions and the fact that every living thing is interconnected (phew). It had taken a team of twenty-eight scientists thousands of hours of calculations and modelling to draw their conclusions. But he explained it super-simply and many things became crystal clear to me. The main point, I realized, and the reason for this book, is that we (*Homo sapiens*) have been terrible friends to the planet. We urgently need to flip that and become a great ally. Otherwise we will be in deep, deep trouble.

Did you know?

Worldwide emissions of CO_2 are close to 40 billion tonnes every year. That's equivalent to 400,000 of the biggest aircraft carriers, if you can imagine those lined up. Next, imagine those vaporizing into the air.[1]

In Chapter 1 we got a sense of the incredible biosphere that we inhabit. We acknowledged some of the brilliant resources held in the Earth's biomes and ecosystems. The truth, as we know, is that we couldn't get by without them. Even more embarrassingly, they could get by quite well without us. (So when you hear people proclaiming they are off to 'save the Earth' it doesn't really work that way, but we know they mean well!) The Earth also does an amazing job at regenerating itself and provides nourishment for us in a variety of ingenious ways every day. But as the Swedish team realized, there are boundaries. I suppose this is a bit like, when you're a small child, you have to be told there are boundaries. If not, you'd just fill your face with candy until you were sick and throw your toys all over the place 24/7. That's what we've been doing with the Earth's resources. We've had no respect whatsoever for the planet's boundaries. We grab, poke, steal and prod the Earth and take too much. We do this so quickly that the poor planet doesn't get enough time to replenish all the resources we snatch away. At the same time, we pour in different sorts of pollution, sometimes literally into the rivers and waterways. I think you'll agree, this is a shocking way to treat a friend. Of course, as ever, the valiant Earth does a great job in trying to neutralize or absorb as much of the chaos as possible. Remember, Earth wants to be a great Goldilocks zone for us where everything is *just right*. But we're pushing the limits so much that the Earth has not been able to replenish, regenerate and recalibrate. Pushing the limits has left us in a crisis situation because we've trashed our own house.

The Swedish sustainability scientists focused on the way that the ecosystems worked together and the resources that Planet Earth provides, and zoned in on the biosphere's ability to regenerate. They crunched the numbers and worked out what resources the Earth could safely provide for us to use, allowing for time to regenerate. Then they divided everything up into a giant chart. We don't have room in this book for a giant chart, but here's a smaller version that gives you an idea.[2]

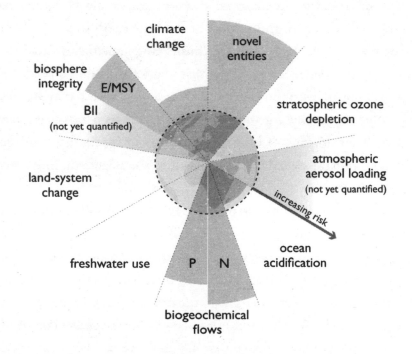

Don't worry if it looks a bit complex. The thing to note is that there are nine different segments. I got the impression that the Swedish scientists would have liked there to be ten, as that's such a great, neat number. But it didn't work out like that. If you've ever played Trivial Pursuit – a high chance, since you like quizzes! – then you can think of this as nine different wedges, dividing up all the resources provided by the Earth. Ordinarily – at least for a period of 10,000 to 12,000 years – the Earth did a fabulous job regulating and replenishing those nine different slices of metaphorical pie. This long and relatively harmonious geological epoch was called the Holocene. Guess who benefited from this period? Yes, it was us – *Homo sapiens*. The Holocene was our time to shine. Unfortunately, rather than being very grateful for this, and taking good care to stay within the boundaries of the Earth, we kept piling on the pressure. Many experts believe that we've been so ghastly to the Earth during the last 150 years in particular that we've destroyed many of the advantages the Holocene has provided. They now think we've pushed the Earth so far that we're not even in the Holocene any more, and must try to live in a whole new epoch. Yes, you read that correctly: pushing, prodding, provoking in every one of these boundaries means that we have become essentially a geological force in our own right. We have done what no gang of humans or any other species has managed to do. We have kicked the Earth, as if it was a football, into a new geological epoch.

And conditions today aren't quite as good as they were before. In fact, unless we take charge and give ourselves a massive talking-to right now, change our ways and respect the Earth and its boundaries, things could get a whole lot worse. The Swedish scientists hypothesize that the more pressure we apply to the different slices of pie, the more danger we are in of pushing the planet's operating systems across dangerous thresholds. Once that happens, we will be in uncharted territory. We simply don't know how the planet will react and what the outcomes will be, but it is very likely that it will make it harder for us to live here, to grow our crops, to enjoy healthy seas and clean air. The type of dangerous, scary 'natural' incidents that you might watch in a disaster film for a bit of a thrill – such as huge fires or floods or hurricanes – will become real-life events.

To avert disaster we need to act right away and there is a very clear place to start: in the bit of the diagram that reads 'climate change'. These days pretty much everyone has heard of climate change, but what the phrase fails to convey is the urgent message: WE NEED TO TACKLE THIS RIGHT NOW! That's why protestors at huge rallies and marches around the world carry placards declaring 'climate crisis' or 'climate emergency' as they agitate for urgent action.

Did you know?

Research from NASA's Jet Propulsion Laboratory suggests that the Earth is putting on weight around its middle. Well, aren't we all? But in this case the spread is caused by melting glaciers.

This whole crisis revolves around an increase of pollutants in the atmosphere. Who would have thought that concentrations of atmospheric gas would ever get such attention, let alone be the thing that would move so many people to come out on the streets and profess their love and respect for the Earth. But that is exactly what has happened as more and more of us come to understand that greenhouse gas emissions are plunging us into a crisis situation. As these gases collect in the atmosphere, they raise the global surface temperature of the Earth and this has huge ramifications.

The greenhouse gas you will probably hear the most about is, of course, carbon dioxide, because this is the primary gas created by our various activities – from keeping warm or cool to growing food. Carbon dioxide is principally caused by burning fossil fuels; in many countries (including mine) we are world-class, Olympic-standard burners of fossil fuel and therefore creators of greenhouse gas emissions. Each year human activities pump about 37 billion metric tonnes of carbon dioxide into the atmosphere.

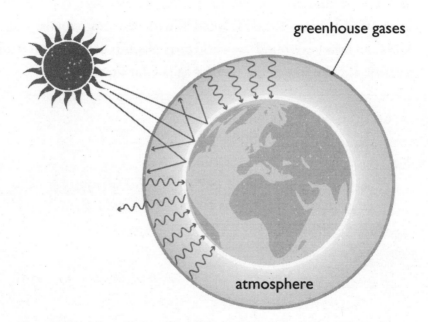

This would be bad enough if it were a recent phenomenon, but in industrialized nations such as the UK, the problem began long before we were born. With the nineteenth century came the famous Industrial Revolution, when people like the engineer Isambard Kingdom Brunel sauntered around in tall hats pointing out where bridges were to be built. At this point, countries in the vanguard of change developed an insatiable thirst for stuff (including bridges).

To produce all of this required energy, and that involved getting coal out of the ground as quickly as possible and burning it.

Since the industrial age we have become ever more skilled at creating greenhouse gas pollution, and that remains very much the problem, especially as it has not been balanced by an equal talent for looking out for the planet. It's this abundance of greenhouse gas emissions that is threatening our very existence, and that of people in countries who have only created a tiny fraction of the emissions that a rich, industrialized nation like the UK is responsible for. As the world warms, a whole heap of unintentional impacts are created, from the melting of the polar regions and the rise of seas, which creates terrible flooding and will render some island nations uninhabitable, to the creation of heat domes, where record-breaking temperatures can fan the flames of wildfires, turning them into catastrophic events. It's enough to leave us pleading with the Earth, 'Can we go back home to the Holocene?!' To which the Earth might reply, 'Only if you urgently stop producing greenhouse gas emissions and work on all the other stuff.' While we're not able to simply return to the previous epoch where everything was 'just right', we can stop the damage we're doing. But we have to act now.

In climate terms, this is often referred to as 'stabilizing the climate' or 'limiting global temperature rise' (remember the emissions cause the greenhouse or blanket effect and those are the things that drive up the temperature). Until recently, most experts thought that we were already locked into disaster temperatures because the huge concentrations of greenhouse gases in today's atmosphere would carry on warming for decades or hundreds of years. But I have rare good news! It now seems that if we go for it, stop burning fossil-fuel resources like coal and gas, and utilize renewable sources of energy instead like the sun and wind (which, by the way, are also free), then we could stabilize the climate within a couple of decades.[3] Isn't that

TWO-STEP PLAN FOR THE PLANET

For those of you who like a to-do list, this one is relatively simple.

1. CRUNCH THE CARBON

The first thing on that to-do list is to stop releasing greenhouse gas emissions at concentrations that are frankly going to kill us, or many of us. The most effective way of doing this is to keep fossil fuels in the ground. The least effective is by carrying on as usual and then hoping that someone clever will invent a machine that will suck dangerous emissions out of the sky.

2. LET THE EARTH DO ITS JOB!

As we saw in Chapter 1, the Earth's ecosystems, from forests to deserts and oceans, all do an incredible job sucking in and storing emissions from our activities. The trouble is we're producing such a great volume of emissions that we're outpacing Earth's natural stores. To use my favourite blanket analogy – we all know what happens when you've got too many blankets on at night: you get hot. In your sleep you have the luxury of kicking the excess blankets off – unless you've made your bed too tightly. The Earth has no such luxury. The blanket traps those greenhouse gas emissions and the world gets warmer. A rising surface temperature might not, at first glance, seem like a big deal – but I assure you it is.

quite the prize? Some might say we won't really deserve a reward, because we've been such a lousy friend to the Earth. But we are being offered this chance, this opportunity and this huge prize. I think it is worth a shot, don't you?

It seems to me that every day we find out more about the Earth and how resilient it could be, if only we shift focus and stop doing things in the same old tired, self-centred and frankly unfriendly ways. It's time for a fresh start and a fresh approach. To do that, I think it pays to understand the climate crisis a little bit more. Not so that we can get demoralized, but so that we can spot and take advantage of real opportunities to join the transition to a fossil-fuel-free future. All of which brings us to our second quiz round.

NOW TEST THE BOUNDARIES OF YOUR KNOWLEDGE
YOUR ESSENTIAL TEN-QUESTION QUIZ

1. As we've established, after 10,000 to 12,000 years we are in effect moving home, and not just across the street but into completely new, uncharted territory that is very far away. We are moving from the Holocene epoch, where thing were *just right* not just for *Homo sapiens* but for all life on Earth, to a less stable, less predictable epoch. But what is that epoch called?

    A. The Anthropocene
    B. The Crucible
    C. The Anthracite epoch
    D. The Forcing epoch

2. We know that humankind has been generating the primary source of global warming, carbon dioxide emissions, since the Industrial Revolution. But according to research by Oxfam and the Stockholm Environment Centre, in the years from 1990 to 2015, annual global carbon emissions changed by what percentage?

A. They dropped by 20 per cent
B. They rose by 60 per cent
C. They rose by 90 per cent on pre-industrial levels
D. They rose by 20 per cent

3. What is the magic number that the global climate regime (that's the top-table stuff organized by the UN, bringing together almost every nation on Earth) is trying to limit global temperature rise to? Many businesses are also aiming for this target too.

A. 1.5 degrees Celsius
B. 2.7 degrees Celsius
C. Zero
D. 3.4 degrees Celsius

4. Let's hear it for the tireless work of climate scientists! After all, we wouldn't understand any of this were it not for the decades of research and data capture in far-flung parts of the globe. It was at the Mauna Loa mountain-top observatory in Hawaii that scientist Charles David Keeling first started measuring atmospheric CO_2 consistently (in 1958). Where do you think the CO_2 level – measured in parts per million – stood back then?

A. 280 parts per million (ppm)
B. 400 ppm
C. 316 ppm
D. 325 ppm

5. We're already experiencing some of the effects of global temperature rise due to climate change. In the summer of 2021 as wildfires and floods linked to global heating ravaged continents, turning familiar places into disaster zones, the Intergovernmental Panel on Climate Change, a group of global climate scientists, released a report that hit the headlines. The UN Secretary General, António Guterres, said the report showed:

A. 'the greatest threat to our existence in our short history on this planet'

B. '[that] the Earth is in a death spiral. It will take radical action to save us'

C. '[that] the climate crisis has already been solved. We already have the facts and solutions. All we have to do is wake up and change'

D. 'code red for humanity'

6. Carbon dioxide is, of course, not the only greenhouse gas emission. Methane is a global warming gas that is twenty-eight to thirty-four times more potent than carbon dioxide. But which of the following is responsible for emitting the most methane?

A. Cow farts

B. Cow burps

C. Cow manure in slurry pits

D. Cows on pasture

7. We produce a lot of global warming pollution that the Earth valiantly tries to remove and store. We know trees are good at this (we've even got a whole chapter on trees coming up because they are so impressive at storing carbon), but which of the following is also a seriously super sucker?

A. Krill, the crustacean that weighs around 1 gram but removes 12 billion tonnes of CO_2 from the Earth's atmosphere each year

B. Rose bushes. In Grasse alone, the perfume capital of France, rose bushes remove a gigaton of carbon, trapping carbon between the dense petals and storing the carbon in complex root systems

C. Bears and wolves reintroduced into forested environments are estimated to be sequestering (capturing and storing) 5.5 billion metric tonnes of carbon (the same as US emissions in a single year)

D. Peridotite rocks in desert areas such as Oman sequester 11 billion tonnes of CO_2 every year

8.

We know that a rise in emissions means a rise in global temperature. In September 2021, when temperatures hit 18 degrees Celsius above average in some parts of Greenland, what happened that caught the scientists at the summit station unawares?

A. It rained for the first time on the Greenland icecap

B. The snow got thicker and scientists were snowed in

C. The Greenland icecap generated solar electricity for the first time

D. Helicopter supplies had been brought in by kayaks

9.

In September 2021 at the Youth4Climate summit, the Swedish youth climate leader Greta Thunberg called out empty rhetoric from 'so-called' leaders, invoking a three-word phrase several times. What was that phrase?

A. 'Wah wah wah'

B. 'Yeah yeah yeah'

C. 'Eat, rave, repeat'

D. 'Blah blah blah'

10.

How alert and engaged do you think we (the general population) really are when it comes to the climate crisis? Some media organizations such as Albert – the sustainable arm of the British Academy for Film and Television (BAFTA) – make it their business to track mentions of 'climate', 'carbon' and 'emissions' in the media. The results make fascinating reading. Which of the following data from UK TV is *not* true?

A. In 2018 'Climate Change' achieved fewer mentions on British TV than 'Cats', 'Cake' and 'Picnic'

B. In 2019 'Climate Change' lost out to 'Dogs' and 'Weddings' in terms of mentions on British TV

C. In 2020 'Climate Change' was over ten times more common on British TV than the term 'biodiversity' and fifteen times more common in terms of a subject than 'banana bread'

D. In 2020 'Climate Change' got more mentions than either 'dogs' or 'cake'

. . . .

ANSWERS

1. The time that we're living in is increasingly referred to as the Anthropocene, so A is the correct answer. But I should tell you that some scientists disagree that there is enough evidence to declare a new epoch, so it remains an informal title. If you haven't heard it before, you may have worked it out from the Greek term for human, *anthropos* and new, *cene*. Used for the first time in a scientific paper in 2002,[4] it reflects the fact that we humans have become the most influential species on the planet. Isn't that amazing (and not in a good way)? One of the weaknesses, I think, with labelling this epoch the Anthropocene is that it might suggest that every living person is equally responsible for causing the pressure on Earth that has caused this shift into a new geological age. But, as we know, that is not true. Someone born into a poor household in Bangladesh has a tiny carbon footprint compared to me. But I do think this label is useful in showing just how destructive a force those of us in industrial nations have become, and how out of step with the entire history of the Earth we are. Never before have humans exerted such influence on the life-sustaining systems of the planet. So part of our task in becoming the ultimate friend of the Earth if we live in a consumerist society is reducing this influence as much as possible.

2. It's B, emissions rose by 60 per cent during this quarter of a century as shown by joint research from Oxfam and the Stockholm Environment Centre, released in 2020.[5] When this research was published it shocked many people. I think we are always shocked when we realize how much of the change now experienced by the Earth has happened on our watch. If you answered A – that emissions had dropped – I only wish you were correct! To become the generation to end climate change, we are going to have to work extremely hard and pull every lever and push every button, but hey, that's what we're here for, right?

3. The answer is A: 1.5 degrees Celsius. This is the temperature that the science tells us gives us the greatest chance of limiting the sort of feedback loops and unstable impacts that will make parts of our precious planet unliveable for some species and communities. In particular we should always be mindful of those who live in low-lying countries, namely Pacific Islanders, who will lose their homes if sea rise is too great. Many are understandably nervous that the 1.5 target is disappearing from our grasp (research in 2021 showed that we are already experiencing warming of around 1.1 degrees Celsius). But we need to remain ambitious and determined. In short, where the Earth is concerned, we have to keep pushing for as close to 1.5 degrees limiting of global temperatures as possible.

4. The right answer is C, the CO_2 level stood at 316 parts per million (ppm), just a little higher than the pre-industrial level of 280 ppm (answer A). At the end of the last ice age, this is where the concentration of greenhouse gases in the atmosphere would have stood. And then … humans invented machines that both required them to burn fossil fuel and made coal mining easier (in his 2015 book, *A Rough Ride to the Future*, the eminent scientist James Lovelock, who we met in Chapter 1, singles out the Newcomen engine invented by Thomas

Newcomen in 1712 as being *the* breakthrough machine).[6] Fast-forward to now: on the day I compiled this question, the CO_2 level stood at 413 ppm, up from 411 ppm the year before. So that's definitely going in the wrong direction! In fact, we passed the 400 ppm threshold (answer B) on 2 May 2014. This was the first time carbon dioxide had touched 400 ppm in at least 800,000 years.[7] To really get to grips with CO_2 levels, check today's level against the level on the day you were born. It's an eye-opener: www.co2.earth/daily-co2.

5. The answer is D, UN Secretary General António Guterres described the report as 'code red for humanity'. In fact, this was the sixth assessment the IPCC had produced, but it was the one that made the headlines. Why? Because it pointed out – in dispassionate language while clearly setting out the evidence – that we humans are on a collision course with our own planet. Perhaps you remember the day it came out? I do. It was not a proud day to be human, but for me it reinforced the idea that *surely we can do better than this*. Well done if you got this one right. The other quotes are all highly motivating too. A is from the US actor and activist Mark Rufallo, B from the British activist and journalist George Monbiot, and C from Greta Thunberg.

6. Apologies for the indelicate nature of this question but it is answer B, cow burps! Many will have said farts (which, let's face it, get a lot of attention), but a cow belch has a higher global warming potential score. This is because belching involves 'enteric fermentation', the digestive process where sugars are converted into molecules for absorption into the bloodstream. Methane is produced as a by-product. Next we have the huge ponds or lagoons where cow manure is processed – they produce a lot of methane. Then finally, farting, where a small amount of methane is produced in the cow's large intestine and then (ahem) let out.[8] So that's burp, pond, fart. Got it?

7. The answer is A, krill! If you answered D, there is a lot of research going into the possibility of storing CO_2 in peridotite rocks in Oman through a process called Direct Air Capture. But it is not happening yet. While answer B is nonsense (I find no evidence that rose bushes have superior carbon sequestration ability), there is some truth in C. A 2017 study from California's Stanford University found that a forest rich in mammals would store more carbon than a forest devoid of mammals. The study, which was based on readings carried out by an indigenous community in the Amazon, found that where there were mammals feeding and living, they increased the efficiency of the environment. It estimated that maintaining mammal populations in the Amazon (in many parts of the Amazon biome, mammal populations are in decline) would help to store 5.5 billion metric tonnes of carbon[9] – equivalent to the total emissions from the US in 2017.[10]

    But let's focus on krill. If you got the right answer here, you understood two things: first, never underestimate the power of small (and perhaps less prepossessing) creatures in nature. You'll be learning by now that it's not all about the big charismatic species. When these small crustaceans, weighing just 1 gram each, come together, they form a powerful carbon force. The second thing you'll have grasped, unless of course it was just a lucky guess, is that krill are not just food for penguins or whales. There's more on krill in the next chapter, by the way, because I'm a huge fan.

8. The answer is A, it rained for the first time instead of snowing. This was so unexpected that scientists at the summit station had no instruments to measure the rainfall (I doubt they had an umbrella either). I always take any opportunity to talk to researchers who spend time at summit stations. I am often struck by how resourceful and adaptable they have had to become in order to keep pace with the rate of ice melt and glacier retreat (due to global heating). When I asked a

scientist from the British Antarctic Survey to describe his job to me once, he said: 'These days I make robots that can swim.'

9. The answer is D, 'Blah blah blah', which Greta Thunberg has used several times since and which has become a shorthand to describe those in authority talking in a nonsensical way about climate action but failing to walk the talk. In that speech, she added it to examples of empty promises and distinctive phrases that had recently been used by politicians, especially speeches given by the UK prime minister, Boris Johnson, in the run-up to the COP26 climate negotiations in Glasgow in 2021. In one speech that went viral, Johnson seemed to refer to those inclined towards eco-friendly behaviours as 'bunnyhuggers'. So Thunberg referenced 'Bunnyhuggers blah blah blah', calling out Johnson's tendency to use strange, dismissive language about action on climate. But she also invoked the long period of inaction from all leaders: 'Thirty years of blah blah blah.'[11] This did a great job of highlighting the gap between rhetoric and action to cut emissions sufficiently in order to limit global temperature rise to 1.5 degrees Celsius (above pre-industrial times). If you answered B you might be remembering the Beatles song 'She Loves You', which is an altogether different vibe.

10. The answer is D: the 2020 results for Albert's sustainability subtitles survey, released in 2021, showed that 'Climate change' mentions were slightly down overall from the year before at around 12,700 mentions across the year. This means that for the third year (since the survey began), 'Climate change' was resoundingly beaten by 'cake' (eleven times more popular) but pulverized by 'dogs' (twenty-three times more popular).[12] The Albert scheme literally scans subtitles across

the UK television networks to produce the data sets for the survey. Outside this work, Albert certifies many British TV shows, including those that attract big audiences such as *EastEnders* and *Coronation Street* – popular soap operas. These shows have been encouraged to slash their carbon emissions in production, through changes such as running studios on renewable energy (backstage). But this is yet to filter through into programming onscreen. Perhaps when 'climate' gets anywhere near to the number of mentions attracted by 'dogs' (and I say this as the proud mother of two rescue pups) we'll know we've had a breakthrough in terms of airtime for Earth.

You did it! That was a tough chapter. You grappled with the Holocene, you wrestled down carbon jargon and you picked up some specifics about how to scale down emissions. Those are all seriously useful skills to have under your belt. And if your score doesn't reflect your enthusiasm, don't forget you can always return later for another run.

NATURAL-BORN HEROES

So here we are already in Chapter 3, and well on our way to repairing our fractured relationship with Planet Earth. We're making good progress, and I want to remind you at this point that you don't have to do this alone. Imagine the backup you have when you can call on the 1.9 million (or even 1 trillion according to some estimations) species of flora and fauna (which is just a fancy phrase for plants and animals) recognized by science and housed in its biomes. In a sense, these are your housemates. Don't make them angry and don't steal their breakfast cereal. Now, for those people who have a particularly anthropocentric mindset, you may have to sit them down and gently break it to them that we share this planet with a lot of other life. The Earth is crammed full of living organisms, all manner of flora and fauna. They are dependent on each other. After all, fauna (animals) can't prepare their own food – have you ever seen an armadillo buying groceries? So they need to feed off the flora and other fauna. In turn, we are dependent on both for all manner of things. When they flourish, we flourish. When flora and fauna populations are abundant, that's a sign of good planetary health.

Did you know?

If you have access to the internet and you love nature, the Encyclopedia of Life is a free online database (eol.org) documenting all 1.9 million species on Earth recognized by science. You will never be bored again!

This chapter (and the next) is a bit like the Oscars for nature and biodiversity, which actually mean the same thing. I like to use 'nature' because, really, when did anyone ever feel emotional when hearing the word 'biodiversity'? If I ask you, 'Do you love biodiversity?', you would probably say, 'Erm, not really.' But if I ask, 'Do you love nature and want to do everything in your power to protect and respect it?', well, you'll likely say, 'Hell yeah!' As you know, that's the kind of reaction we're looking for in this book. Being a great friend to the Earth obviously means being a great champion for all of nature. It means bigging up the smallest creatures and understanding what important ecosystem services they perform. Take earthworms for example. They are the recipient of tonight's first major award. Why? Well, think of the service they perform, converting organic debris to feed the soil, aerating the soil as they go. You couldn't buy a service like that. But how often do you think about earthworms? How much do you know about them? Probably not often and not much. Yet there are at least 6,000 known species[1] and you can get more than 1 million in an acre of soil. The jury (I'm back to my pretend Oscar ceremony) was bowled over by the earthworm's dedication and unsung skills. (By the way, I was brought up in a Devon village where we had an annual 'worm charming festival', so I probably think about worms more than the average person.)

We heard briefly about the incredible carbon-storage capabilities of krill, aka the super-shrimp, in the last chapter. But let's go a little deeper, because it turns out krill poo goes deeper than we thought too (and be thankful that I didn't call at least one chapter in this book 'Cow Farts and Whale Poo', because I definitely thought about it). The humble krill is the recipient of our second major award this evening. The jury was impressed by the way krill swarm in the ocean, eating phytoplankton and then excreting carbon and nutrients (including iron) that fertilize the oceans in the same way that we fertilize fields on lands to keep them productive.[2] Then there's the poo. Formed in relatively large pellets, these drop to the ocean floor, safely storing carbon. Other research shows that these creatures are also more courageous than first imagined. For decades scientists thought that krill just floated about in the top 150 metres of Antarctica's waters, which would make sense frankly, because this is where they have easy access to phytoplankton. But no, it transpires that they go deeper than that. An underwater camera found krill to be swimming around just above the seabed. This matters because the lower they are in the water column, the less chance their poo pellets have of being swept away and the greater the chance of them settling on the seabed and storing carbon.

Unlike we humans, who like to be rewarded for our good endeavours, I suspect krill and earthworms don't really need a lot of praise. But I tell you what they do need. A bit of RESPECT. Because at the moment, instead of getting rewarded, they are being pushed into the danger zone, again by our everyday actions. How do we repay krill for being a critical part of a process that removes

12 billion tonnes of CO_2 from the atmosphere every year? Yes, you've guessed it: we increasingly overfish them. Krill are in high demand to feed livestock and even for use in pet food. I don't think these are worthy uses for the unsung carbon heroes of the Southern Ocean, do you?

In the age of the Anthropocene, it's not easy to be a small species. It's time we paid more attention to the tiny creatures that seem abundant and are at the bottom of the food chain. For example, the great biologist E. O. Wilson, known as 'Darwin's natural heir', spent many years studying ant colonies.

Did you know?

According to the WWF, 'Scientists have a better understanding of how many stars there are in the galaxy than how many species there are on Earth.'[3]

As well as krill, earthworms are another small creature that scientists are worried about. Earthworms, so fundamental to the health of our soils, and therefore our food supply, are impacted by climate change as they are particularly susceptible to wet weather, and with climate change comes more rainfall. Earthworms have been found for the first time ever in the northernmost forests of Canada – it's not that they are not welcome, but they shouldn't be there. The thawing of the previously frozen soil thanks to climate change, together with fossil-fuel pipelines and development, have helped to spread them north. Once in these forests, their chomping of organic matter, always so helpful elsewhere, causes a problem: they are releasing extra carbon.[4] An earthworm in the wrong place turns from a friend of the Earth to a foe.

In fact, it is not easy being *any* creature in the Anthropocene. All across the globe, species of plants and animals are being lost at a faster rate than ever before. Around the world, a quarter of species of plants and animal groups are vulnerable to extinction.[5] As I was sitting down to compile this book, I read a piece online that told me that twenty-two species of birds, fish, mussels and bats (alongside one species of plant) were declared officially extinct in the US during 2021. They're gone. They're never coming back. That is the largest number of extinction declarations ever in a single year. I was pretty shocked. I wasn't even aware that there was an ivory-billed woodpecker (although now, having seen a picture of it, I do mourn its loss even more). But this is just a small fraction in an era that has been dubbed the Sixth Great Extinction. Of course, there is absolutely nothing 'great' about it.

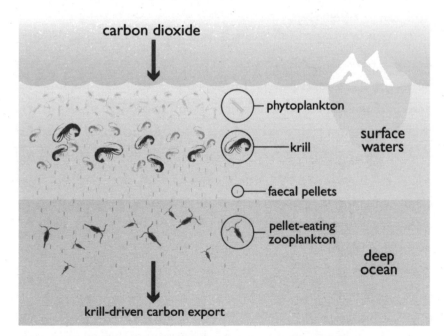

Tiny krill are big players in the ocean carbon cycle.

When I read a shocking bit of nature news like that, I always look to see what laws and targets have been agreed by global nations to stop such destruction. I found out that a list of twenty targets was agreed in 2010 by the UN. They covered specific ecosystems such as coral reefs but also practical changes in the way our human system works against nature, such as removing government subsidies that damage nature. These became known as the Aichi Biodiversity Targets (named after the prefecture in Japan that hosted the summit). I checked to see what progress had been made on the Aichi Targets, and found to my horror that none had been met at the time of writing. Now, bear in mind that these were established in 2010. In fact, the world has never met a single UN target to halt or limit the destruction of the natural world. This failure seems unacceptable to me. We simply cannot afford to lose species, even the ones we don't know of and are yet to discover. We are just custodians of this planet, and it is part of our job to keep it habitable for all the species that we still haven't discovered.

Incredibly, we've only realized this fairly recently. In 1980 a team of research scientists were startled by the immense diversity of insects in tropical forests. In one famous study of just nineteen trees in Panama, 80 per cent of the 1,200 beetle species discovered were previously unknown to science. Of course, the sad fact is that the rate of species loss is now so great that it seems inevitable we will never know of many new species. But I think we should be more determined than depressed by that thought. And I believe there are things we can do right now to help. Because being a true friend at this point in history means being an activist where possible to hold those in power to account. The best way of doing this is to be as well versed as possible in nature and to join and sign up to as many organizations that safeguard nature as possible. We cannot rest for one minute. If you live in an elected democracy you need to make very sure that your elected officials know without any shadow of a doubt that this means everything to you. Whether it's combating local habitat destruction for development, or the loss of a species thousands of miles away driven by changes in land use thanks to the things that we consume (we'll talk more about this in Chapter 5), you have a role to play and a voice to raise in order to speak out for nature. Because defending nature is defending the Earth. The two go hand in hand.

So then, now that we're all fired up about defending nature, let's test your existing knowledge.

WILL YOU MAKE THE SURVIVAL OF THE FITTEST?

YOUR ESSENTIAL TEN-QUESTION QUIZ

1. I want to start with perhaps one of the most intriguing and accessible animals: birds. Of the few positives resulting from the global pandemic and associated lockdowns many experienced, one was hearing how millions of people discovered birdwatching, a hobby that I really encourage any prospective ultimate friend of the Earth to get into. But how many bird species brighten up the planet?

A. 4 million
B. Around 250,000
C. 10,000
D. 1 billion

2. Is the following true or false? There are more tigers kept in the US (where tourism drives tiger breeding) than there are wild tigers in Asia and tigers in zoos worldwide.

3. Earlier in this chapter, we met some earthworms in the wrong place. In 1995, which unwelcome visitor arrived on Fregate Island in the Seychelles?

A. The caecilian, a limbless amphibian
B. The armoured spider, *bib arme* (it's actually a big beetle)
C. The rat
D. The whip scorpion

4. OK, let's go for human versus animal now. Here are four notoriously speedy swimmers; three are fish, one is Olympic swimmer Michael Phelps. I want you to put them in order, from slowest to fastest recorded speed in the water.

A. Great White; Shortfin Mako; Michael Phelps; Atlantic Bluefin Tuna
B. Michael Phelps; Atlantic Bluefin Tuna; Great White; Shortfin Mako
C. Michael Phelps; Shortfin Mako; Atlantic Bluefin Tuna; Great White
D. Atlantic Bluefin Tuna; Shortfin Mako; Great White; Michael Phelps

5. Halfway through this round, and I'd like to focus on the planet's toughest creature. This one has earned the prefix 'indestructible'. It's the one animal, more than any other on Earth, that seems to be resistant to almost everything. But what is it?

A. The koala
B. The bee fly
C. The wood frog
D. The tardigrade (also known as the water bear)

6. In December 2021 wildlife officials in Florida, US, began a special rescue operation involving hand-feeding manatees, the huge aquatic creatures known as 'sea cows'. This was deemed necessary to save them from starvation, a knock-on effect of huge algal blooms from polluted waters. But what did wildlife officials elect to feed the hungry manatees?

A. Imported seagrass
B. Romaine lettuce
C. Orange pulp
D. Almond paste

7.

In the decade 2000 to 2010, how many rhinos were estimated to have been killed in the Kruger National Park at the hands of poachers?

A. 10
B. 400
C. 4,000
D. 40,000

8.

Which country currently holds the record for the number of amphibians?

A. Costa Rica
B. Bhutan
C. Australia
D. Indonesia

9.

As we lift our gaze from our own navels, we start to notice some intriguing things about other species. Some creatures live much shorter or longer lives than *Homo sapiens*. Talking of timespans, how long can a bowhead whale live for?

A. Over 200 years
B. Into their eighties
C. Into their mid-forties
D. 1,000 years

10.

Our final question for this round tests your knowledge of data gathering and observation – a very important part of ecology and conservation science. Experts warn that we are experiencing a huge drop-off in insect numbers (it's not good news to be low on pollinators for the obvious reason of their importance in growing our food!). But how do they measure this 'insecticide'?

A. By monitoring air currents and placing insect traps on hedgerows

B. Through a splatometer on a car windshield

C. By running a survey on hedgehog populations (the two are interlinked)

D. By spraying insecticide into a tree canopy and counting the resulting shower of dead insects

. . . .

ANSWERS

1. The answer is C, $10,000^6$ and 389 of those species are flagged as being vulnerable to extinction due to climate change. Now, you might have heard an estimate of double that because in 2016 a team of scientists decided to define 'species' in a way that defied convention. They defined it as any group with a unique, shared set of traits regardless of whether it could mate with other taxa (species).[7] This meant an explosion in bird numbers, up to 18,000 bird taxa.

   One thing you'll need if you're into ornithology is an authoritative bird book (or these days a good app). I still use the one my grandfather gave me for my eighth birthday! But imagine being the first to produce a field guide, before photography was available. Take a bow, John James Audubon, the Haitian-born Frenchman (he later became an American citizen) who drew 489 bird species in 1,065 life-size illustrations. It took him twelve years to complete.

2. Sadly this is true. It's unclear exactly how many tigers there are in the US, but it is thought to be well in excess of 5,000. This compares with under 4,000 remaining wild tigers in Asia. Those in the US are mostly captive tigers, privately owned and living in people's backyards, roadside attractions and private breeding facilities. The number of tigers in zoos worldwide is calculated at 1,659.[8]

3. It's C, the rat, to the consternation of the island's conservationists. The beetle and the whip scorpion were casualties, with the beetle population dropping by 80 per cent as rodents chomped through them. If you guessed A, you couldn't be more incorrect! While the rat is an invasive species to the Seychelles (a non-native that flourishes to the detriment of native species), the caecilian is known as a 'deep endemic': caecilians are thought to have been one of the creatures that made the crossing from Gondwana to the Seychelles, so have a very long lineage.

4. The answer is B, with respective fastest speeds being Michael Phelps 4.3 mph; Atlantic Bluefin 25 mph; Great White 30 mph; Shortfin Mako 35 mph.[9] No offence to the incredible Michael Phelps, a twenty-eight-time Olympic medallist, but it's worth noting now and again that humans are outpaced by other species.

5. The answer is D, the water bear (tardigrade). These 'marvels' of biology comprise at least 1,300 species that live in moist homes, particularly in a film of water on lichens and mosses and moist soils. They are truly incredible, evading all sorts of adverse conditions that would kill off most other creatures. This includes my personal favourite 'skill' of being able to

dehydrate to a sort of desiccated state (like dried coconut) when their homes get dried out. But the sad news is that, according to research by Danish scientists, when that dry period goes on for a longer period – and in a warmer world this is what is likely to happen – the water bear becomes as vulnerable as the next microscopic moss dweller. It's curtains! Another reason to join the global battle to halt climate emissions. Do it for the tiny moss-dwelling water bear.

If you chose C, wood frog, I'm afraid frogs are likely to be one of the creatures hit hardest by climate change. Ditto with koalas, another of the top ten species likely to be worst affected.

6. The answer is B, romaine lettuce.[10] Increasingly, conservation is becoming emergency conservation. When you're trying to avert disaster in this way, you have to minimize the risk and, based on previous experience of trying to rescue manatees, Florida wildlife officials decided it's the best choice – providing some nutritional value and hydration, and being least likely to cause negative health impacts. Given that manatees need to eat 10 to 15 per cent of their bodyweight a day to flourish, you can see that the rescuers had to get hold of a lot of replacement food, and lettuce can be sourced from local farms. To feed a manatee for a day takes about 50 pounds of lettuce, enough for 100 Caesar salads.[11]

But the question you probably have is: 'Why are the iconic, plump and laid-back manatees starving so that their ribs show and local people describe seeing them as "heartbreaking"?' The answer is to do with pollution and timing. In the winter manatees return to warm-water feeding grounds to feed on seagrass. But in 2021 the seagrass beds were devastated by algal blooms. Such blooms are caused by pollution and lack of oxygen in the water, killing off the seagrass. In 2021 this has led to a record-breaking year of manatee deaths. Well done to the wildlife officials for stepping up, but how we wish that wasn't necessary.

7. The answer is C, 4,000 rhinos according to estimates shared with *National Geographic* magazine in 2021.[12] If you guessed D, you'll be shocked to learn that the total wild population of rhinos is estimated to be 27,000.[13] However, in the war against poaching there is some better news. In 2018, hounds imported from Texas and fitted with GPS trackers began to be deployed to track poachers (who were tracking rhinos). This method has been proven to be pretty successful, leading to a 24-per-cent drop in poaching in the park.

8. The answer is A, Costa Rica, which positively excels as a home for frogs. The most comprehensive survey on frog wealth puts it this way: 'Costa Rica exhibits the greatest species richness per unit area in Middle America, with a total of 215 species reported to date. However, this number is likely an underestimate due to the presence of many unexplored areas that are difficult to access.'[14] But the country's natural wealth does not end there. It is a biodiversity hot spot and if you happen to be from there, you have an excellent head start when it comes to being an excellent friend to the planet. Costa Rica's achievements include axing the country's military budget, diverting those funds towards education and the environment, and protecting 25 per cent of the country's land as parks and protected reserves in which logging and deforestation are not permitted. If you're Costa Rican, we're pretty jealous. Or rather I mean, we're going to try and be a bit more Costa Rican ...

9. The answer is A: over 200 years. We know this because during a bowhead necropsy (the animal version of an autopsy), scientists determined the approximate age of the creature by carbon-dating weapons used – in this case old harpoon points embedded in the animal's body – in earlier attacks on it that the whale survived. A memorable case in 2007 found a harpoon point used in Victorian times, suggesting that the whale was at least 130 years old. Contrast such a lifespan with the rapid pace of habitat change in the

Arctic, which is warming twice as fast as any region on Earth. What does that mean for the wildlife populations?

10. The answer is B, a splatometer. I know it sounds like something a circus clown would employ, but insect researchers are deadly serious. In Europe, data on the huge decline in insect populations has been gathered in Denmark, Kent in the UK and parts of rural France – all regions that rely on pollinators to grow food. The splatometer involves placing a grid over part of the number plate or windscreen and driving round as insects hit the car. Afterwards researchers simply count the number of splatted insects. In rural Denmark, data collected using this method between 1997 and 2017 found an 80-per-cent decline in insect numbers.[15] In Kent, a splatometer survey in 2019 found a 50-per-cent decline from 2004.[16] Researchers even use 'vintage' cars to make sure that they are replicating conditions exactly. If you answered D, spraying insecticide into a canopy, this was the method employed in the Panama survey that resulted in 1,200 beetle species being counted, referenced earlier in this chapter.[17] I'm not sure it sounds a very Earth-friendly method though!

Whatever your score, I hope this chapter served as a reminder of how awesome fellow living creatures can be. While they are natural friends of the planet in a variety of different ways, they don't have a voice and it's up to us to protect their habitats. In fact, safeguarding habitats for other species is a key part of the job description when it comes to being the ultimate friend of the Earth, which makes this chapter something of a milestone. Congratulations on reaching this point, you are well on the way!

4

HUG A TREE, MARRY A MYCONAUT

Let's begin with trees. In some cultures (i.e. mine in the West) and in the recent past, if you wanted to insult someone you might call them a 'tree-hugger'. This would imply that they were weirdly obsessed with nature, disconnected from 'real life' and often (deeply inferred) that they were in need of a shower or bath. As we teeter towards the ecological abyss, these are all characteristics that I think are commendable (we waste a lot of water in industrialized countries on power showers, which by the way can use more water than a bath). Besides, the more we learn about trees the more we should frankly want to throw our arms around them. They're completely brilliant. Being in their company and close to them has also been shown to be very good for our mental health. So next time someone calls you a tree-hugger, smile your biggest toothiest grin and say 'Me?! Oh thank you so much.' It's the greatest compliment on Earth right now.

In the dry jargon of climate politics and green finance, trees are 'valued' for the great 'ecosystem services' they offer. These terms relate to what you may have always thought was great about trees, such as the way they provide a home to a huge range of wildlife and the way they filter water and stop erosion. They are also increasingly being talked up as a nature-based climate solution, which is a fancy way of saying that they are very good at sucking up and storing carbon. Such an emphasis is very much to be encouraged if it stops humans from chopping them down, but we have to be careful not to become overly reliant on this.

Did you know?

There are an estimated 3,000,000,000,000 (3 trillion) trees on Planet Earth. That means roughly 420 trees for every one of us. Your job is to keep your 420 alive and flourishing, and then double the number!

While it's true that trees are super suckers, absorbing 16 billion metric tonnes of CO_2 per year as they grow and storing it in their roots, they also release CO_2 when they are felled or if they die (as they decompose, CO_2 is released back into the atmosphere).[1] And far too many are being felled or lost through fire or disease or extreme weather events. It sounds like we have a lot of trees, but there are probably half as many as there should be. OK, you might say, let's just plant loads more. That's also important, but there is no silver bullet here because it's mature established trees that bring most of the benefits. The number-one way of stabilizing our planet and securing its health is to stop burning fossil fuels and protect the trees we do have, particularly in old-growth forests.

Did you know?

A mature tree absorbs carbon dioxide at a rate of 48 pounds per year. In one year, an acre of forest can absorb twice the CO_2 produced by the average car's annual mileage.[2]

Now, sorry to rain on your parade (again), but we haven't been doing well here. Even our famous trees and most cherished forests are struggling. Take the Amazon rainforest, which I think we can all agree is pretty famous and important. Well, in 2021 we received a warning when scientists confirmed that the Amazon rainforest was now emitting more carbon dioxide than it was able to absorb. Thanks to pressure from land clearing for beef and soy production, hotter temperatures and droughts (driven by climate change), the Amazon has passed a dangerous threshold, turning from a carbon sink to an emitter. As burning forests produces three times more emissions than the forest can absorb, the conclusion would seem glaringly obvious: stop clear-cutting and burning forests *immediately*! Instead, in some areas rates of deforestation are continuing to soar, which makes it all the more vital to support global efforts to end deforestation. The good news is that the global push behind those efforts gets stronger every day. Our understanding of forests is also flourishing. When I was at primary school, I always remember drawing trees on their own.

I would hope now that I would be encouraged to draw them collectively, as a forest. It would have helped me to think of them as interconnected communities that communicate with each other, rather than lonesome, oversized plants.

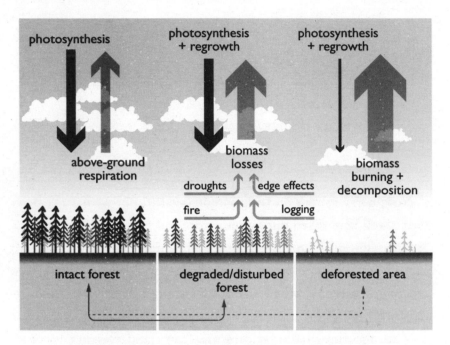

Intact forests are the best for regulating the biosphere.

Thrillingly, researchers have found this runs far deeper than we imagined. It is time for us to factor in not just the root system of trees but the network of fungi roots – known as hypha – and the extraordinary influence these have. We are only just beginning to get to grips with the glories of fungi and the complex interactions that take place below ground. But underneath your feet are hidden networks where fungi facilitate communication and the swapping of information and nutrients between plants and trees; they are assisting and maximizing the forest's work as a carbon sink. It's not just fascinating, it is potentially life-saving, and this story is only going to get bigger and bigger as we make more discoveries about the way that fungi work.

Did you know?

There are as many as 6 million species of fungi, 90 per cent of which haven't yet been described. (It's a complicated and painstaking process, from collecting a new species, proving it is indeed 'new', through to having the species accepted by a formal natural history body!)

At the moment mushrooms are often touted for their potential to replace 'bad' materials with high environmental impact – such as plastics and livestock products – with fully biodegradable, low-carbon materials. We've begun to hear about them in consumer goods and packaging products and even in construction and engineering materials. The cool thing about them is that they can essentially grow as much material as you need. But there's much more to come because fungi appear to support life on Earth in all manner of incredible ways and form enormous networks, seemingly communicating with each other.

Research so far suggests that some of the most important networks are found in old-growth forests, where fungi grow on trees over 400 years old. Yet these are the forests that are most at risk not only from logging but also from wildfires. To lose these forests means losing the fungi as well, before we've even fully worked out the extent of their importance. This cannot be allowed to happen, because experts are crystal clear on two points: (1) there is no future without forests, and (2) our future must also be about fungi.

Are you ready to test your knowledge and potentially find out more? Here are your questions for this chapter.

CONVINCE US YOU'RE A TOP-FLIGHT MONKEY-PUZZLER
YOUR ESSENTIAL TEN-QUESTION QUIZ

1. Let's start off on a sobering note. Approximately how many trees are cut down every year?

A. 1 trillion
B. 15.3 billion
C. 7.7 billion
D. 500 million

2. Which country is the largest forestry country in the world, i.e. boasts the most trees?

A. United States
B. United Kingdom
C. Russia
D. Brazil

3. We already know that trees are one of our most important carbon sinks, but *which* trees, and in what context does research suggest they are the real super suckers?

A. Pine trees grown over at least 100 hectares

B. An existing mixed tropical forest[3]

C. 100 different species of saplings newly planted in urban areas

D. Plantations that border oceans

4. On 29 August 2005 Hurricane Katrina made landfall. Ripping through the US Gulf Coast it uprooted or seriously damaged more than 320 million large trees. According to research, how much carbon did this release into the atmosphere?

A. Emissions equivalent to those produced from flights by the global airline industry in 2019

B. Emissions equivalent to the amount absorbed by all the trees in the US over one year in 2005

C. Emissions equivalent to a rocket launch

D. Equivalent emissions to those produced by 900,000 petrol cars

5. What is the name for someone who professionally reads tree rings?

A. An epidemiologist

B. An arboriculturist

C. A dendrochronologist

D. A horologist

6. Analysis in 2019 showed that the Earth lost a football-pitch-worth of pristine tropical forest every six seconds.[4] But in which country was the biggest surge in tropical forest loss?

    A. Bolivia
    B. The Congo
    C. Brazil
    D. Costa Rica

7. Let's take two questions now on tree research from two eminent scientists. In 2017 research led by the Chinese American climatologist Rong Fu showed trees in the Amazon to be capable of what?

    A. Looking after their offspring
    B. Acting as the lungs of the planet
    C. Communicating profusely
    D. Creating their own rain clouds

8. What is the name given to the biggest and oldest tree in the Amazon rainforest by celebrated Canadian ecologist Suzanne Simard?

    A. The General
    B. The Mother Tree
    C. The Survivor Tree
    D. Hyperion

9. What has the symbiotic network of trees and other plants beneath our feet, made possible by fungi, become known as?

    A. The Wood Wide Web
    B. The Agaric Underpass
    C. The Hypha Highway
    D. The Fungal Fairway

10.

In October 2021, Mylo was unveiled to an audience in Paris. But who or what was Mylo?

A. The first electric car with an interior derived from mushrooms

B. A robot that can be deployed on oil spills, using fungi to capture hydrocarbons

C. The world's first mushroom leather to be made into a bag for a major designer and feature in a runway show

D. A new composite fibre made of mushroom and polypropylene strong enough to use on space rockets

· · · · · · · · · · · · · ·

ANSWERS

1. The correct answer is B: approximately 15.3 billion trees are cut down each year.[5] There are numerous threats to trees, including extreme weather events and disease (all connected to climate change), but the human axe and chainsaw remain the biggest threat to them. Most of us, of course, are indirectly responsible for the felling of trees due to our consumption habits. Use of forestry-derived products, ranging from loo roll to fashion items (an increasing amount of clothing fibres are derived from wood pulp) might seem incidental, but if we don't research and take care to buy only certified sustainable products then we may unwittingly be helping to bankroll catastrophic forestation. Talk about death by a thousand paper cuts.

2. The answer is C, Russia, with a reported 642 billion trees,[6] containing over one-fifth of all the world's forest and containing 111 billion cubic metres of wood.[7] Then comes Canada and

Brazil, with the US at number four in the league of largest forestry countries with 224 billion.[8] While all trees are important, our biggest concerns are rightly for tropical rainforest, aka primary forest, which is the most biodiverse and locks in the most carbon. Environmentalists may sometimes seem to be overly preoccupied with what happens in the Brazilian Amazon, but it actually makes good sense to focus here.

In my own country, the UK, trees are often accorded special importance, especially those of historic significance such as the Major Oak in Sherwood Forest, Britain's largest oak tree (folklore suggests Robin Hood and his Merry Men took shelter in its canopy). But in truth we have relatively low tree cover overall. Tree cover stands at about 13.2 per cent of land. Much of this forest cover lies in Scotland in the north. If we look at England itself, forest cover plummets to around 10 per cent. This is lower than South Sudan, one of the countries that has the lowest coverage in the world.

3. The answer is B. An intact, species-rich forest sequesters twice as much carbon as a planted monoculture,[9] which is why plans like A Global Deal for Nature (the companion pact to the Paris Agreement) emphasize the need to formally protect existing forests and to keep existing codes in place and strengthen them. Safeguarding our trees is not just a nice thing to do, or the right thing to do, or a tree-hugger thing to do – it is the smart thing to do to ensure our survival.

4. The answer is B. According to a 2007 research paper by Tulane ecologist Jeff Chambers, around 100 million tonnes of CO_2 would have been produced by the mass tree loss caused by Katrina, equivalent to the annual net amount sequestered by US forestry.[10] If that seems like a terrible irony, it gets a bit worse. When trees are taken out like this, it leaves fewer trees to photosynthesize and they decompose, releasing the carbon they have stored up to that point. They turn from sinks

to emitters, releasing large amounts of carbon dioxide. This boost in carbon dioxide levels in the atmosphere increases the likelihood of extreme weather events, which in turn increases the likelihood of more trees being destroyed. In other words, tree-felling hurricanes beget tree-felling hurricanes. If you answered A, you may be interested to know that worldwide flights from this period totalled eight to nine times more than this amount.[11] If you answered C, you were a little way off. Launching a rocket would be much less, at 300 tonnes per launch.[12] But if that sounds like a bargain – don't even think about it! It's still a lot of carbon that we cannot afford.

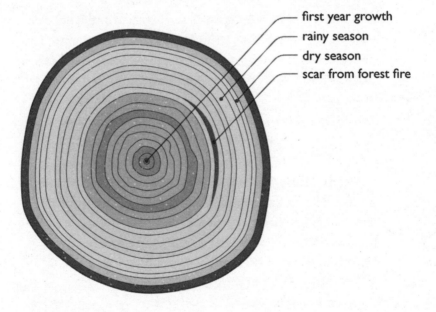

first year growth
rainy season
dry season
scar from forest fire

5. The answer is C: dendrochronology is the science of dating tree rings. What a fascinating life that must be. Each of those rings gives a snapshot of conditions on Earth at the time it was formed in the tree's trunk. Think of each ring as a time capsule giving information on the climate, tree growth rates and the chemistry at that time. Scientific advances in our understanding

of how trees network reveal that some of their messages seem to be encoded and passed on in DNA, giving older trees a form of memory (see the work of Dr Suzanne Simard). That's another reason why it is so important to save older established trees; they are the linchpins of the forest.

6. The answer is A, Bolivia. All in all, the data showed that the tropics lost 11.9 million hectares of tree cover in 2019, one of the highest years of forest loss on record. This scale of tree-cover loss is especially heartbreaking given that tropical primary forest is considered the most important type of forest for sustaining biodiversity. Brazil accounted for the biggest overall loss of humid tropical forest loss, but the ignominious title for biggest surge in destruction went to Bolivia. Here there was an 80-per-cent leap in loss of tree cover compared to previous years. Experts pointed out that this surge came after government encouragement to expand agriculture, leading to greater clear-cutting and burning of forests and an increase in fires. Meanwhile, the United Nations estimates that over 100,000 acres of rainforests are being destroyed each day. Unless we take decisive action to protect them, the last remaining rainforests could be consumed in less than forty years.[13] Yikes.

7. The correct answer is D. Professor Fu's research showed that trees in the Amazon are able to create light rain clouds and rain showers. This answers the question as to why the rainy season begins two to three months earlier in the Amazon than elsewhere in the region. The process of transpiration causes rain clouds, which create their own circulation large enough to shift wind patterns, blowing in more moisture than the ocean.[14] These clouds bring rainfall in other areas, crucially to croplands in Brazil's so-called Bread Basket.

    Professor Fu's further research shows how the atmosphere is becoming thirstier and this is driving the catastrophic wildfires that have become a feature of modern life in many regions. These, in turn, release more carbon and destroy wildlife and homes. Has the clearing and cutting of tropical rainforest ever seemed more idiotic?

(By the way, if you were tempted by any of the other answers, they all contain a kernel of truth, but of course there is only one right answer.)

8. It's B, the Mother Tree. Suzanne Simard's research has changed the way we think about trees. She describes Mother Trees as being the biggest and oldest in the forest and the glue that holds the ecosystem together. They are canopied cradles of biodiversity, utilizing their huge canopies and abilities to photosynthesize to provide food for the whole web of life round them. She has also hypothesized that they possess a form of memory and intelligence through which they are able to help forests recover from disturbances.[15] Find out more at Dr Simard's Mother Tree Project: mothertreeproject.org.

9. The correct answer is A, the Wood Wide Web. Geddit?! In common with almost every aspect of fungal greatness, this area is not yet fully understood and needs a lot more research. But what fungi specialists (mycologists) do know is that this vast network fungal network allows flora and fungi to share a whole load of resources and information. For example, it enables trees to supply seedlings with sugars. Meanwhile, trees that are ailing and perhaps dying can dump food resources into the network, where they can be conveyed to healthier neighbouring flora. The best way to think of it is as a lot of chatter, negotiating and planning going on beneath your feet. The network even seems to allow trees and plants under threat to spread a warning through

chemical signals. Honestly, weekend walks will never be the same again now that you know all this is going on.

10. The correct answer is C. Mylo is a mushroom alternative to leather used in the Frayme Mylo bag, part of iconic designer Stella McCartney's summer 2022 collection and shown at Paris Fashion Week in October 2021 (that's how the seasons work, in case you're confused). Mylo is made from mycelium and developed by Bolt Threads, the Californian materials company that has worked closely with McCartney for a number of years. Bolt scientists claim they are able to reproduce what happens on the forest floor, creating Mylo from mulch, air and water and 100 per cent renewable energy.

   These innovations have spawned a new area of material science and could help to displace demand for livestock leathers linked to environmental destruction in the Amazon in particular. They are also notably less polluting than 'vegan' leathers that are made from plastic synthetics derived from fossil fuel. After all, mycelium is infinitely renewable.[16]

So, how did you do? Watch out later in the book for some more mentions both of trees and of the wonder of the mycelium network, and of how you can agitate to protect and grow knowledge of them. One thing is for sure: the fungi/mycelium (as the network is known) is just going to grow as a story. It could become the biggest topic in ecology, which brings us back to the cheeky title I chose for this chapter. Myconauts, who specialize in mycology (the study of fungi), could be the next rock stars! But for now, it's time to meet another invaluable climate and nature ally, another bona fide rock star of the biosphere.

IT'S THE OCEAN, STUPID

You might be surprised by the abrupt title to this chapter. How rude, you might think! Especially as we're probably feeling a bit smug by this point in the book. After all, haven't we just raced through a whole load of facts and snippets full of promise and potential to help us restore the Earth's health? We have, and I don't wish to dismiss the excellent work so far. But while we've met some incredible carbon sinks and natural restorers that influence the atmosphere, the chemistry, the weather and a host of conditions on Earth, they are not the biggest player.

That honour goes to the ocean, an Earth regulator of almost unfathomable importance. As has been said by many, there's nothing quite like it in the universe. Yet for most of human existence we have failed to pay the ocean the respect it's due. Instead we just think of it as something to be moved across, through or beneath, conquered and exploited, principally through fishing and mining. This is not a good idea. The fact is that without the ocean, there's simply no liveable planet. For the purposes of this book and this mission, you cannot be the ultimate friend of the Earth without extending the hand of serious friendship to the ocean.

Did you know?

The ocean acts as a carbon sink for at least a quarter of the atmospheric CO_2 emitted by human activity.[1] Its incredible circulation system made up of currents – aka the global conveyer belt – helps keep temperatures within the right range for life and it produces over half of the world's oxygen.[2]

By the way, you might think it's strange that I speak of one global ocean in this way. I know someone who owns his own flashy boat and really took exception to this. After all, he has a skipper's certificate. 'There are five oceans,' he says (with some validation), 'the Atlantic, Pacific, Indian, Arctic and Southern Antarctic.' Yes, there are five named oceans and some prefer to count seven (specifying the North and South Atlantics and North and South Pacifics). However, I noticed that the great oceanographers and marine scientists who I most admired talked of the ocean as a single thing. Some even referred to it as a liquid bridge between land. This is a deliberate attempt to refocus us on the fact that ours is primarily an aquatic planet. In *Ocean: A Global Odyssey*, a huge book of discovery by one of my heroes, Dr Sylvia Earl pointedly includes the quote from NASA scientist Christopher McKay: 'If some alien called me up [and said], "Hello, this is Alpha, and we want to know what kind of life you have"[3] – I'd say, water based ... Earth organisms figure out how to make do without almost anything else. The single non-negotiable thing life requires is water. So take that, Captain!'

The ocean covers more than 70 per cent of the Earth and holds 96.5 per cent of all the world's water. It's incredible to me that we don't spend 96.5 per cent of every day talking about it and considering its needs. But guess what? Yes, most of our attention is taken up with what happens underneath our own two feet. Since we spend most of our time on terra firma, we think that's where the action is, but we're *wrong*!

Did you know?

The ocean is under intense threat from human-caused climate change. Between 2009 and 2018, 14 per cent of the world's corals were killed, according to a 2020 report by the Global Coral Reef Monitoring Project.[4]

The ocean is a critical carbon sink for us, but it also regulates climate and the Earth's chemistry in a way that we have only recently come to understand – which also explains why we were a bit late to the ocean party. It also serves as a vital carbon sink through ecosystems including coral reefs, seagrass meadows, kelp beds, salt marshes and mangroves. The ocean is estimated to hold fifty times more carbon than the atmosphere.[5] That makes it a gold-star AAA-plus-rated nature-based carbon solution, although on the flip side, damaging the ocean ecosystem increases carbon emissions. But still, many experts think humans have been tardy to translate the emerging evidence of the pivotal importance of the ocean to a liveable Earth.

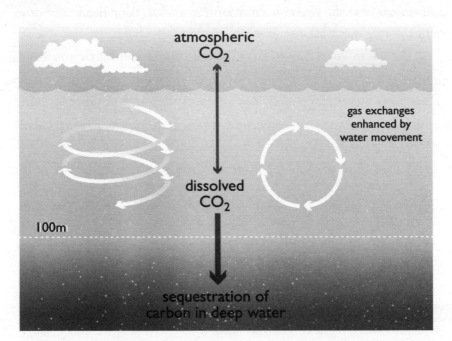

How the ocean works as a carbon sink.

It is only very recently that we've seen the ocean acknowledged as a nature-based climate solution and brought into the global climate processes of COPs and other fancy climate negotiations. Small wonder that 'It's the ocean, stupid!' is something oceanographers and other marine scientists must have spent years muttering to themselves, especially at these big conferences and meetings. Yes, it's time for all of us to step up and pay the ocean the respect it deserves.

Thinking differently about the ocean should lead you to make some different decisions. We have a lot of bad habits when it comes to the ocean, often treating it like a great convenience store from which we can take whatever we like. Assumptions and habits like this mean that we unthinkingly trash the ocean from the projects we bankroll (via our savings and pensions) to the seafood we buy. (Tip:

try swapping out 'seafood' and replacing it in your head with 'sea life' – it gives you a very different perspective!)

Did you know?

More than 80 per cent of the ocean has never been mapped, explored or even *seen* by humans.[6]

In any case, we can't keep on as we are. Just as on land, we are biting the hand that feeds us. Our activities, from industrialized fishing to deep-sea mining, threaten to destroy this incredible, complex, interconnected system that we rely on, before we've even grasped just how amazing it is. There is good news, however: we know much more than we did! This is the era of breakthrough discoveries about how the ocean works and what lives in it. Thanks to all sorts of strange bits of equipment and engineering and scientific innovation, such as submersibles that are able to take courageous researchers to explore new depths below, the discoveries are coming thick and fast. But will they be fast enough? Well, that partly depends on us. Even if you live hundreds of kilometres away from the nearest coast and have traditionally had little interest in the sea, the ocean needs your voice. As ever, that starts with testing how much you know and spotlighting where there's room for improvement. Here are your ten questions from the Deep Blue.

PLUMB THE DEPTHS OF YOUR KNOWLEDGE
YOUR ESSENTIAL TEN-QUESTION QUIZ

1. In 2019 the journal *Science* published a piece, 'A New Narrative for the Ocean'. Which of the below is taken from it?

A. 'The ocean is not too big to fail, nor is it too big to fix. It is too big to ignore'

B. 'What's happening in the ocean is biological annihilation'

C. 'The ocean's bottom is at least as important to us as the moon's behind'

D. 'We do not inherit the ocean from our ancestors, we borrow it from our children'

2. Which nation became the first on Earth fully to protect its sharks, banning all commercial shark fishing and establishing a shark sanctuary?

A. The Maldives

B. Australia

C. Indonesia

D. Palau

3.

One quarter of all fish and seafood is caught using trawlers. Each year an area of the ocean is trawled equivalent to which of the following?

A. The size of France

B. The continent of Africa

C. The Amazon rainforest

D. South America

4.

Of course it's not just fish that humans take from the ocean. We take large volumes of sand for concrete, and gravel, and as the competition hots up for rare earth metals many mining companies are eyeing up the ocean floor for these too. What is the rather polite name given to this oceanic resource-dash?

A. The Blue Gold Rush

B. The Blue Boom

C. The Global Sea Grab

D. The Blue Acceleration

5.

Staying with the uncomfortable theme of pillaging our beautiful ocean, what is the name given to the lumps of rock on the seabed that are causing excitement in the mining industry?

A. Ocean nuggets

B. Polymetallic nodules

C. Sharktooth coal

D. Rare ocean metal

6. Increasingly, we are standing up to ocean exploitation in a variety of ways. In 2021, a documentary film taking aim at the fishing and seafood industry built a social media audience of millions and stormed the Netflix ratings. What was it called?

A. *Blood Shrimp*
B. *Seaspiracy*
C. *Ghost Fishing*
D. *Massacre of the Oceans*

7. OK, so we get the message. The oceans are in serious trouble! Now let's talk about what we are doing to protect it. A Global Alliance led by the UN is pushing for how much of the ocean to be protected by 2030?

A. 3 per cent
B. 30 per cent
C. 90 per cent
D. 50 per cent

8. In 2021, a new-found coral reef, stretching 2 miles off the coast of Tahiti in the South Pacific, was discovered by researchers. How did those who made the discovery describe the 'pristine' coral?

A. As 'underwater fungi'
B. As resembling 'roses'
C. Like 'miniature sombreros'
D. Like the skirt of a 'whirling dervish'

9. Not surprisingly, the ocean holds the deepest point from the Earth's surface. But in which of the following can the deepest spot be found?

A. Puerto Rico Trench
B. Philippine Trench

C. Mariana Trench

D. Japan Trench

10. It took ten years to complete 540 marine expeditions involving more than 2,700 scientists from eighty-plus countries, and by 2010 humanity had been gifted its first Census of Marine Life. Which of the following species was NOT catalogued in the first marine census?

A. Marine deepjelly

B. Large dumbo octopod

C. The yellow porous *Quadris braccis*

D. The 600-year-old tube worm

. . . .

ANSWERS

1. The answer is A, 'The ocean is not too big to fail, nor is it too big to fix. It is too big to ignore.' For me this quote encapsulates everything. The piece by two renowned marine ecologists, Jane Lubchenco and Steven D. Gaines, called on the world to come together to make sure that the ocean didn't fall into further decline, and to realize and restore its vibrancy and health, not least for the part of the global population who rely on it for their sole source of protein. It marked a real stepping-up by agencies across the world to address the threats to the ocean and seems to have spurred the growth of many small volunteer-led programmes worldwide.

2. The answer is D, Palau, a nation that inhabits an archipelago made up of around 340 islands across the Pacific, which banned all commercial shark fishing across its exclusive

economic zone in 2009.[7] It's no secret that, globally, sharks are experiencing a huge decline in numbers; overfishing threatens one third of all sharks and rays.[8] Palau's example proved successful and divers were soon reporting an increase in shark sightings as shark tourism (not fishing) soared. In 2015 Palau went further still, designating 80 per cent of this zone as a no-take zone, and banning all fishing and mining.[9] In just ten years, sharks and other populations have increased by up to five times the level pre-ban. This is what restoration looks like and it raises the question: why are we waiting to protect more areas of the ocean? Let's go!

3. The answer is C, the Amazon rainforest. According to a major study (published in *Nature* magazine) the area of the ocean floor trawled each year is around 5 million square kilometres.[10] The Amazon rainforest covers approximately the same area.[11] The significance of this is often played down, given that it represents 1.4 per cent of the total seabed, but that is to underestimate, and perhaps intentionally minimize, the havoc wreaked by bottom trawling: heavy nets are dragged along the sea floor scraping up anything in their wake. Again, to me, this speaks of our lack of interest or regard for the ocean. Would we stand for bulldozers being sent into wild areas of land to pursue tiny creatures, churning up and smashing every bit of life as they went about it. I don't think we would. So why do we allow it for the oceans?

4. The answer is D, the Blue Acceleration, used to describe the period we are entering. As resources become compromised on land, I suppose we should have expected we would turn our attention to the ocean. Harvesting ocean resources is hardly new – the expedition boat HMS *Challenger* set off from Great Britain in 1872 to check out various ocean deposits – but the scale of need and aspiration is very different today, and some would say very troubling. This is a

story of pressure, and all the signs show us that the ocean needs less pressure, not more. We are already learning that the decline of ocean health is currently outpacing human global action.

5. The answer is B, polymetallic nodules. Described elsewhere as resembling lumps of charcoal or coal, they are much more interesting than that. The nodules grow at about 1 centimetre every million years on the seabed by absorbing metal compound in seawater, forming around material on the ocean floor, such as a clam shell or shark's tooth.[12] There are potentially billions of tonnes of polymetallic nodules on the ocean floor, and as they contain the hottest (as in most in demand) of rare earth minerals such as nickel, cobalt and manganese, you can understand why many corporations are desperate to get to them.

    On the plus side, it is claimed this could be the answer to supplying the materials we need to fuel the huge green transition to renewable energy and electric vehicles (after all, the metals for these must come from somewhere!). But at what cost? Biologists and oceanographers, and just those of us who love the ocean, can't help but be very nervous at the mention of robotic bulldozers mining for nodules on the ocean floor, further destroying a fragile ecosystem.

6. You guessed it! Or did you? It's B, *Seaspiracy*, the film made by filmmakers Ali and Lucy Tabrizi that exposed the relentless impact of the global fishing industry on the entire ocean. The film had many detractors. Some NGOs and contributors questioned its accuracy, especially when it comes to the film's unsympathetic portrayal of so-called sustainable fishing. But arguably these are outnumbered by its supporters, who maintain that the film is not about scientific accuracy but about telling the story of the crisis faced by our ocean, which is primarily down to our habit of using it as a larder, with

ever-increasing cost to Planet Earth. If you opted for *Blood Shrimp*, this is a powerful label and concept applied to the global shrimp fishery in the film, tainted by slave labour and ecological destruction.

7. Let's start by saying, if the answer was that the Global Alliance only wanted to protect 3 per cent of the ocean by 2030, I would have to hand in my resignation to Planet Earth. In fact that paltry amount is where we are now. The correct answer is B, 30 per cent by 2030, a bold campaign represented by the hashtag #30by30 or #30x30 (later there are some pointers as to how you can get behind this). If 30 per cent doesn't sound all that ambitious, the strength of this plan lies in the location of the protected areas aimed for. We know that Marine Protected Areas (MPAs) work and that by instilling no-take zones preventing fishing, mining and other activities, wildlife and habitat can regenerate. But so far these zones have tended to be near the coast. What needs to happen is to target the high seas – the harder-to-police areas of the ocean, away from the coastlines. These are the regions most rich in biodiversity, so the plan is that some of the protected areas will include the high seas for the first time.

8. The answer is B: the coral was described as resembling roses, the flowers traditionally associated with romance, beauty and Valentine's Day in many cultures.[13] I'm sure that, as a nature-lover, you'd rather discover this 'pristine-looking' coral reef than receive two dozen roses. This reef was a very special discovery. Globally, coral reefs have been depleted due to overfishing and pollution. Climate change is also harming delicate corals – including those in areas neighbouring the newly discovered reef – with severe bleaching caused by warmer waters. The researchers who discovered this reef remarked that it appeared to be 'pristine' and had

not been damaged by the bleaching event that occurred across most of the South Pacific in 2019. Now we must work to keep it that way!

9. The deepest spot of all is the Challenger Deep, which lies in answer C, the Mariana Trench in the Pacific Ocean. The estimated depth here is around 11,000 metres below the Earth's surface, some 2,000 metres more than the height of Mount Everest.[14] It used to be thought that these depths were effectively dead zones and couldn't support life. But our understanding of the deepest parts of the oceans has been revolutionized thanks to submersibles, craft that are able to descend and take photographs and video footage. Flat fish, large shrimp, huge crustaceans and even a snail fish have been spotted in the Mariana Trench. So, unfortunately, has a plastic bag, showing that plastic pollution really does pervade every part of the Earth.[15]

10. It's C, as this is my tongue-in-cheek description of the popular animated character SpongeBob SquarePants. Apologies for my bad Latin if you happen to be an expert. 'Absorbent', 'yellow' and 'porous' are characteristics of SpongeBob according to the show's theme song. The theme song also reveals that SpongeBob lived in a pineapple. None of these facts made it into the Census of Marine Life (COML) but 6,000 potential new ocean species did. These included the first animal that lives without oxygen. What made this project so important was the discovery that the ocean was more interconnected and richer in variety of life than many had imagined. To borrow from *Hamilton* (yes, the stage show!), how lucky we are to be alive right now. Because while we often focus on the threats and levels of destruction – it would be weird if we didn't – we have both evidence and the motivation to transform the way we view the ocean and the way we treat it. I think that's pretty exciting.

Congratulations on reaching the midway point in your quest to be the ultimate friend of the Earth. I really hope you've got to know the planet a little better and reawakened your curiosity and admiration for the natural world. If you ever find your enthusiasm dipping, take a moment to watch a favourite nature documentary or dip into a nature/climate-themed book. (There are a few suggestions at the end of this book if you're looking for inspiration.)

In the meantime, let's continue our mission. We're going to move now into new territory, trying to understand more fully the extent of our impacts on the Earth and to unravel some of our behaviours and systems. As ever, the goal is the same – we just want to do better by our planet. Let's go!

SUPERCONSUMERS

This is the chapter when many of us get a bit embarrassed. We might even find it challenging to meet the quizzical gaze of Planet Earth. Because in industrialized countries we have a lot of stuff, and much of it is pointless and hard to justify. I've made a concerted effort to improve and consume more carefully but, wow, I still have a lot of stuff. The planet would probably like to know at this stage what exactly we think we're up to with all this consuming. We trawl seabeds for grit and the savannahs for sand – no biome is immune or off limits. To add insult to great injury, almost all of it ends up as pollution! But the point to grasp here is that everything we own and use has a 'footprint' that is made up of the energy and resources that go into creating the product or services in question (this is called embodied energy).

One of the things we need to be really alert to is the greenhouse gas emissions that have been created as a result of making, using and disposing of a product. In addition, every product comes with a 'secret' backpack of resources. This is the hidden waste from the extraction and processing of materials that you never get to see, and if all your stuff comes from producer countries thousands of miles away, most of the hidden waste will be left there. This includes the ore, the rubble, the rare earth minerals, the water, the toxic chemicals used to reveal the ore, the soil … I could go on. Suffice to say, it might be hidden from us, but Planet Earth feels it very keenly.

When you ask people what is the most harmful thing for the planet, very often they will give you an answer such as 'overpopulation' and single out developing countries for putting too much strain on planetary systems by having too many children. But this is not the most harmful thing. Research shows that it's the relentless consumerism of those in wealthier nations that is driving the emissions crisis; every time we purchase something carelessly, in effect we're adding fuel to a very large bonfire. From the buildings we inhabit and the transport we use, to the clothes we buy and toys we play with, an enormous 70 per cent of greenhouse gas emissions we create is related to the production and use of products.[1]

Did you know?

Each year humans strip 100 billion tonnes of resources from the Earth's crust to be made into stuff.

Again, some of us have done more to cause the problem than others. Many fellow citizens of Earth *under-consume*; they don't get enough calories from food and live without access to regular electricity or essential hygiene products. Many citizens of Earth live on the scraps of our waste. This robs them of the chance to create and manufacture their own goods in a way that would provide a sustainable livelihood. This chronic inequity (or unfairness) seems to be a really hard nut to crack; most people don't even like to discuss it. But I think the Earth would like us to. It's going to be a hard conversation to have though, full of tough love. Because it turns out that those of us in very consumerist countries have developed some extremely odd habits and views about 'stuff'.

I've been following this overconsumption story for a while, travelling all over the world from Bangladesh to China and the Brazilian Amazon biome, just trying to find out who pays the true cost of the stuff we feel entitled to. It won't surprise you to find out that the planet and some of its poorest citizens get hit with the bill time and time again. I interview people constantly about their consumer choices, I go through their rubbish bins and wardrobes (usually with their approval) and we talk about where the planet fits in. But because I've done this over many years, and I've been through many rubbish bins and wardrobes, I can tell you that we in 'rich' countries, or consumer countries, have become increasingly odd about stuff. We make more crazy decisions, we are buying faster than ever before and we're doing more mental gymnastics to justify our very poor choices. Yes, drivers of crazy consumption, such as Black Friday (the annual price-drop jamboree), have got a lot to answer for.

I recently interviewed a very nice young man called Josh about his choices in consumption for a radio show. He was admirably truthful about his 'fashion habit'. I was quite amazed to hear him say that every five weeks he threw away his jeans and trainers and replaced them. This, he informed me, was a strategic wardrobe choice. He liked his jeans very tight and after five weeks they tended to fray and split at the seams. 'Has this young man never heard of a needle and thread?' emailed a shocked listener (one of many actually). But although many of our listeners were flabbergasted, among Josh's peer group this was not considered odd behaviour, although Josh conceded it was 'bad for the planet'.

Even worse was to come. I was flicking through a newspaper (I know, how vintage!) and a very strange story leapt out at me that I haven't been able to forget. The story featured a social media influencer pulling a very disgruntled face as she sat in her shiny car, with the door open to reveal her outfit: a jumpsuit. The headline explained why she was looking so peeved. 'Model "gutted" as £18 jumpsuit "ruins" her new £60k Porsche.'[2] It transpired that some of the dye from the cheap-as-chips jumpsuit bought from an online fast-fashion retailer had transferred to the pristine creamy-coloured leather interior of the very expensive car, basically destroying the worth of the vehicle. Whoops!

I think I would be upset too. Perhaps you're thinking that it serves her right, but the point for me was that we now consume so frenetically and are so divorced from the real impacts of the origins of items and the cost of manufacturing them that we have little regard for the trade-offs unless they directly inconvenience us.

Way down the supply chain of fast fashion there's a story about dye houses. There has been a decline in standards and a loss of control. Twenty years ago a fashion brand would have a strict list of dye houses it trusted to carry out this important operation. The quality control and finishing would be set to a high standard. But things have changed as brands produce in greater volume at ever-declining cost. Take the villagers I interviewed ten years ago in garment-producing districts in Bangladesh and India. They didn't need to be told what colour was in fashion in 'rich' countries because their rivers that the dye houses fed into were already dyed that colour – in this case, cerise. For me that is a more serious concern than the upholstery in a luxury car.

As we know, we are now using resources at such a rate that we are outstripping the biosphere's ability to regenerate them. A good analogy here is a credit card that we borrow on too heavily. In effect this has a spiralling effect. If you are not paying off the minimum debt, that debt increases. That's how we're treating the Earth and it is no way to treat a friend. The reason we do so is both complicated and at the same time rather simple. We're trapped in systems of overproduction for overconsumption. To unlock solutions, we're going to have to understand what is going on with us, what drives us to consume in such great quantities. And that involves that most mysterious of organs: the human brain.

Did you know?

Each day 27,000 trees are chopped down so that some humans can have toilet paper!

To understand something about our mindset when we're shopping, neuroscientists have attempted to find out what our brains do when we shop. In 2007 a group of US scientific researchers from the academic institutions of MIT, Stanford and Carnegie Mello scanned the brain activities of fashion shoppers.[3] They discovered a conflict: one part of the brain is stimulated by the pleasure of acquiring something new (and I mean really stimulated, as if you've eaten twelve sacks of sugar) and another processes the pain of the cost of our purchase (you could interpret this as a soothing process). The study also seems to suggest that the buzz you get from fashion shopping comes from the sensation of wanting something (not from owning, wearing or keeping it).

The roots of hyper-consumption stretch way back, and often they are built into the way we move stuff around the Earth. Our trade routes are based on slavery, colonialism and fossil fuels. In 2021, a huge container ship, the *Ever Given*, got wedged in the Suez Canal, blocking the passage of global trade for over a week. This was a rare moment in which we the public actually acknowledged the extraordinary size and scale of global shipping – usually it's just something that happens in the background. Typically, it's only when something goes wrong that we take notice. But if you were to go back to 1869, when the Suez Canal opened, you might notice that at this point we were locked into a fossil-fuel future for global trade. The Suez was built to run from north to south, whereas the prevailing winds in the area run from west to east. This ruled out sailing ships, until then the predominant, yet admittedly rather slow, form of delivering silk for top hats and turtle shell for collars (the sort of cargo I imagine from this period). Instead, the era of coal-powered steam was ushered in and bunkering ports that offered stop-off points to refuel took on a central role in global trade, one that dominates to this day.

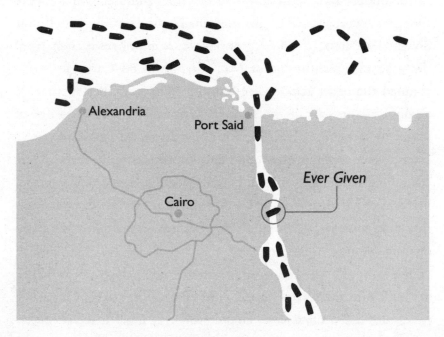

The *Ever Given* container ship blocked the Suez Canal.

Similarly, the beginnings of fast fashion were probably sown as far back as the eighteenth century with the invention of the Spinning Jenny, a machine that sped up cotton-yarn production and enabled the first ready-to-wear garments. By the middle of the nineteenth century, London gentlemen were writing letters to the newspapers complaining that their housemaids had been struck by feckless overconsumption and were squandering their wages on new dresses. (As always, women were painted as the hopeless drivers of destruction.)

The question becomes less about how we got here, and more about what we do next. Because now that we have all this information and evidence that our beautiful Earth is suffering from our drive to consume (and other odd decisions that we make), we can rise to the challenge, cut consumption and change our habits (in Chapter 8: Live Your Best ('Earth-Friendly) Life', we'll navigate some cool ways of doing this). To help in this, it pays to think about how a change of mindset and habit will also benefit us. The truth about owning and accumulating more and more stuff is that it is very tiring. Unless you have a super organizer like Marie Kondo to pop over and help you, you're going to spend a lot of time tidying and storing stuff you don't need. It's not that surprising that new lifestyle movements, whether living in a van (#vanlife) or a tiny house, are all about living with little and valuing experiences over acquisition. The more we can make this the new aspiration, the better for both the planet and our mental health. This sort of outlook is much lighter on the Earth too, and proponents claim it's a far more authentic and equitable way to live, consuming your fair share rather than guzzling everyone else's carbon budget. But before we move on to some strategies concerning how to achieve that transition, we need to get to grips with the size, the scale and the impact of our mountain of consumables. Are you ready for ten questions that get to the heart of stuff and of turbo-charged consumption?

IT'S THE SUPPLY-CHAIN BRAIN STRAIN

YOUR ESSENTIAL TEN-QUESTION QUIZ

1. As we've already discussed, the stuff we own and use is bound to cause greenhouse gas emissions. But how much of our emissions is our consumption believed to be responsible for?

A. 10 per cent. It's negligible compared to the oil and gas industry

B. 90 per cent of emissions is caused by the stuff we produce, use and trash

C. 60 per cent of greenhouse gas emissions and 80 per cent of the world's water use is from stuff consumed by us

D. We don't calculate it like that, as most countries have net zero targets and these emissions aren't included

2. How many new clothes from 'virgin resources' (that means new rather than recycled materials) do you think are produced globally every year?

A. Over 100 billion

B. 8 billion, just over one item per person

C. 3 trillion

D. 100 million

3. OK, now let's think about footwear. How many (pairs of) shoes do you think are made worldwide on an annual basis?

A. 100 billion
B. 3 million
C. 24.2 billion
D. 11 billion

4. Let's stay with today's fashion consumerism, but look at how consumer culture drives trends. What is the name given to the type of video content on social media usually posted by shoppers proudly showing bulk purchasing from fast-fashion retailers?

A. Fashion haul
B. Style dash
C. Bouji TikTok
D. Bargain parlay

5. Now I know we're in the territory where the credible looks incredible, but can you pick out the one answer here that is false?

A. Twice as many copies of Shania Twain's *Come on Over* album in CD format have been sold than there are trees on the planet
B. The number of Lego people on Planet Earth now outnumbers actual human people. In 2015 a Lego person was produced by the Danish manufacturer every 3.9 seconds
C. At any point in time there are more than 6,000 ships carrying containers full of stuff around the world. According to the World Shipping Council, a total of 1,382 containers are lost at sea each year
D. Between 2006 and 2018, fights and stampedes during discount events for Black Friday – the global day of bargains and sales of consumer goods – caused the deaths of 11 shoppers and serious injury to 109 people

6. The **400-metre-long container ship, the *Ever Given*, got stuck in the Suez Canal in 2021. How many barrels of bunker oil does the shipping industry consume per day according to industry sources?**

A. 30,000 barrels of bunker oil. This is less than it was five years ago, as it now runs increasingly on hydrogen

B. 5.5 million barrels of bunker oil for a fleet of around 70,000 ships

C. 1 billion barrels of bunker oil for the entire global fleet

D. 10,000 barrels of bunker oil; the rest is ethanol

7. **Some materials have more cachet than others. We know 'diamonds are forever' because the marketing company De Beers told us in the 1940s! They are also the hardest natural substance on Earth; the only thing that can scratch a diamond's surface is another diamond. What scale determines this?**

A. The Martindale Test

B. The Kimberley Process

C. The Tungsten Test

D. The Mohs Scale

8. **In September 2019, online retailer Amazon revealed its total carbon footprint for the first time. It disclosed that during 2018 all operations led to emissions equivalent to burning 600,000 tankers of oil (44.4 million metric tonnes of carbon dioxide).[4] Amazon CEO, Jeff Bezos, subsequently launched the $10 billion Bezos Earth Fund to help fight climate change. But who said this: 'The people of Earth need to know: when is Amazon going to stop helping oil and gas companies ravage Earth with still more oil and gas wells? When is Amazon going to stop funding climate-denying think-tanks like the Competitive Enterprise Institute and climate-delaying policy?'[5]**

A. The Fridays for the Future Movement (aka the School Strikes movement)
B. Amazon employees
C. Bette Midler
D. Erin Brockovich

9. **Which consumer electronic device is responsible for creating the most emissions?**

A. Fridges
B. Plasma screen TVs
C. Smartphones
D. Kettles

10. **The American documentary filmmaker Lauren Greenfield, who is well known for chronicling the most conspicuous consumption on the planet, has described a new 'dangerous' version of the American dream, one that has 'morphed from an attainable goal, the result of hard work, to a fantasy way of life characterized by self-indulgence, celebrity and narcissism'. What is the name she gives to this phenomenon?**

A. The Versailles Effect
B. The Kardashian Effect
C. Schema Distortion syndrome
D. Generation Wealth syndrome

. . . .

ANSWERS

1. It's C: a massive 60 per cent of GHG emissions are caused by consumers according to the *Journal of Industrial Ecology*. But it is also true that some net zero targets (such as the UK's) do

not include this massive proportion of emissions because most manufacturing of consumer goods takes place in developing nations, who must therefore account for the emissions there. Is that fair?

2. The answer here is A, 100 billion[6] and rising. By 2050 the equivalent of 500 billion extra t-shirts is projected to be added to this total. That's pretty scary when you think that already around 20 per cent of the global fashion industry produce goes unsold and needs to be shredded, burned or dumped. What a waste! If you opted for the '3 trillion' answer, you might have been confused by the total worth of the global fashion industry, which is estimated at a cool $3 trillion.

3. The answer is C, 24.2 billion,[7] enough in theory for three pairs each. But, of course, we don't divide the spoils equally. In the US an average of seven new pairs of shoes per person entered US wardrobes in 2018. Have a think about your own collection. What are they made from? Unless you go for a very traditional shoe – say a completely leather brogue with a lace, or a very 'ethical' shoe such as a sandal made from a former car tyre (yes these exist!), the chances are that your shoe is made of lots of different materials glued and stitched together. If you wear sneakers/trainers/runners (pick as appropriate), these are likely to include a lot of plastic fibres. You probably know what I'm going to say – it's hard and it's complicated to pick apart different materials. Therefore it's hard to recycle shoes. This is a problem, because we have so many of them.

4. It's A, the fashion haul! Shoppers parade their haul to bedroom cameras with a breathless enthusiasm before instructing viewers to 'download the app to shop the links'. Across the world there have been attempts to push back against using social media to drive unsustainable consumption. These include initiatives such as campaign group Fashion Revolution's Haulternative (where haulers post videos showing sustainable

and pre-loved merchandise) and Love Not Landfill, where fashion lovers focus on increasing longevity.

5. OK, there is one false answer here and it's … A, the one concerning Shania Twain. Shania's 1997 album did sell heavily, making the list of top-ten-selling albums of all time, and that does impress us much (that's a Shania joke). However, even that is still not a match for the 3 trillion-plus trees that are on the planet. (And remember, this is not enough trees! We need more.)

6. It's B, 5.5 million barrels a day, which is over 50 per cent of the total global fuel oil demand. Meanwhile, bunker fuel is also described as a 'heavy marine oil' and more casually as being 'as thick as molasses'. You get the picture! This is a dirty fossil fuel, giving global shipping a huge footprint – shipping is thought to be responsible for 2.5 per cent of all greenhouse gas emissions. Can the shipping industry reform in time? Some fleets are switching to gas (still a fossil fuel, but arguably with a lighter footprint) and some are looking to be fossil-fuel-free in the next few years. Some innovators are looking to old-fashioned ways of traversing the ocean and developing huge sails to allow at least part of the journey of the cargo ship to be completely fossil-free, using good old wind power.

7. If you picked D, the Mohs Scale, then you get the glittering honours here! The full name is the Mohs Scale of Hardness, named after nineteenth-century mineralogist Friedrich Mohs. Diamonds score a 10, and quartz a mere 7.[8] In recent times, the international diamond trade has come under scrutiny for human rights violations, including allegations of so-called blood or conflict diamonds (the latter is a UN term for any diamond mined in areas controlled by forces opposed to the legitimate ruling authority). The Kimberley Process, established in 2003 and bringing the diamond industry together with NGOs, including the UN, is a way of providing assurances to

consumers that diamonds are from a 'clean' supply chain free of conflict. If you're in the market for a diamond, make sure you look for this accreditation.

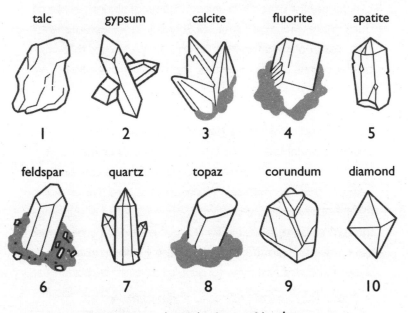

Minerals on the Mohs Scale of hardness.

8. The answer is B, Amazon employees. During this period a group of employees known as Amazon Employees for Climate Justice lobbied Amazon not only to address its own emissions, but also to stop providing computing services to oil and gas companies (Amazon argued that all companies were entitled to the same services).[9] The AECJ was widely credited with pushing the company towards some big moves on behalf of the Earth – Amazon's Climate Pledge and the $10 billion Bezos Earth Fund. In 2021 two climate justice activists and tech workers who had been instrumental in the creation of the AECJ, Maren Costa and Emily Cunningham, settled with Amazon.[10]

They claimed they were wrongly fired for their activism. They remain climate justice activists to this day, and we salute them! Erin Brockovich wasn't involved this time as far as we know, but look out for a question on her famous activism in a later round. Bette Midler caused a kerfuffle on social media when she tweeted in support of the Amazon activist-employees, so you're forgiven if you plumped for C, but sorry, still no point!

9. The answer is C, the smartphone. It is estimated that 3.5 billion people now use smartphones, so there's a high chance you have one next to you right now. Although these are obviously small devices and don't take that much power to charge, their high footprint comes from all their ingredients. This is because most smartphones require sixteen of the seventeen rare earth minerals; substances such as neodymium used to create very small but powerful magnets in our phones. On top of that, building every phone requires the extraction of materials such as cobalt, gold, lithium (for the batteries) and silicon (for the processors). Then add on lead and tin for processing. You can see where I'm going with this – these are all materials scattered across the globe and difficult to reach (although actually not that 'rare' after all). We know that every time we extract materials from the Earth, that causes pollution. To add insult to injury, on average we upgrade our phones every two years (some people do this far more often) and fail to recycle the old model even though it contains such important resources.

10. The answer is B, the Kardashian Effect. Although I'm a bit of a sucker for reality television I have somehow managed to avoid the Kardashians in the main. I sometimes feel I'm the only one left on this planet who has something of a phobia about them. However, I feel the Earth shudders every time they launch a tsunami of copycat consumerism on social media, which they dominate.

If you're even less au fait with their work than me, let me explain. Over the years the Kardashian–Jenners have become the most famous family in the US. They began with an initial reality series, *Keeping Up With the Kardashians*, and now dominate both the reality space and social media. Their connection to consumer goods is difficult to overstate. Each member of the family is involved in launching and promoting fashion and cosmetic products as well as an extravagant luxury lifestyle. While fans tell me the Kardashians often promote sustainability, such as refusing single-use plastic water bottles, to me this feels like a drop in the ocean compared to their relentless plugging of more stuff. Answers A and C are decoys, as they are both a nod to Lauren Greenfield's brilliant previous films, *The Queen of Versailles* and *Generation Wealth* – cataloguing the same phenomenon.

So that's it, the end of our sixth quiz. I wonder how you got on. Did you remain as clueless as the 1990s film of the same name that celebrates conspicuous consumption (I think that applies if you scored 3 or less). Or is there a glimmer of hope that you can free yourself from the bad habits and attitudes that see you upgrading your smartphone and craving diamonds (score 3–6). If you scored between 6 and 8 you are definitely starting to put the planet first. If your score came in over 8, it's your time to shine.

We are now over halfway through our bid to become Earth's ultimate friend and you're starting to display true class. But for now, we've got a landfill dump to deal with and it's not looking pretty.

7

FROM WASTERS TO WARRIORS

It's a truth universally acknowledged that all of the stuff we met in the previous chapter needs to go somewhere. Sadly for the Earth, that 'somewhere' means dumps, holes in the ground or huge bonfires. Yes, one of the things humans seem to excel at is trash-creation. In our hands the useful becomes useless in seconds. Just who is taking the brunt of this diabolical talent? Well, you know the answer to that: it's the poor old planet. Not only are clothes, technological gadgets, children's toys, lampshades and a host of other items being produced faster and in greater numbers than ever before (see Chapter 5), but we are throwing them away at the fastest rate in history.

Along with our proficiency in creating carbon emissions, creating rubbish is another of the worst things to excel at. As we've seen, not only does it mean that we're squandering the Earth's precious resources, but by trashing products too early and in a chaotic way – dumping them in landfills, burning them in incinerators, or shipping them to other countries and making them someone else's problem, we're losing the opportunity to reclaim and reuse them. I could cry for those lost resources! Meanwhile, when we landfill or incinerate or dump in the ocean or waterways, we also create more pollution, both atmospheric and chemical – such as the tiny fragments of plastic that now infest the Earth's waterways. This is pretty dumb given we are in a climate and nature crisis. If the definition of madness is doing the same thing over and over again and expecting a different outcome (attributed to Einstein, as is just about every brilliant quote – they can't all be his, can they?) then the way we approach waste is definitely due a rethink.

Luckily, as we'll touch on later in this book, around the world your fellow global citizens are reimagining waste in all sorts of incredible and innovative ways. In fact, they're designing away the very concept of waste. It's often pointed out that there is no waste in nature (not for the first time, I'm reminded that whenever we're looking for inspiration, we should just look directly to the Earth). I'm conscious we have a lot on our to-do list given that we need to be the generation that ends climate change *and* deforestation *and* inequality. But we also definitely need to be the generation that dumps waste as a concept, never mind in its physical form. We simply can't afford to create much more of it. We are at full capacity for rubbish on Planet Earth. You might think that sounds overly dramatic, but it's not just a question of the waste produced on a daily basis – especially in industrialized countries. It's also the waste we have produced historically that is still with us. One of the most durable forms of waste materials, arguably outside nuclear waste, is also one of the most common forms: plastic!

Plastic waste is now everywhere. That's official. Since the 1950s, when things really got going, humans have generated 8.3 billion tonnes of plastic. The 1950s may seem like ancient history to you, but when you remember that the Earth is considered by most scientists to be around 4.7 billion years old (give or take 50 million years),[1] the age of plastic is but a blink of the eye. Look, though, at how much damage it has caused. We are drowning in plastic. The total mass of plastics on the planet is now over twice the mass of all living mammals.[2]

Did you know?

So far, humans have generated 1 billion elephants' worth of plastic (in weight) and most of it is still hanging around the planet!

Just 9 per cent of plastic has been recycled. Nearly 80 per cent remains with us in some format and in recent years we've come to realize that a huge amount of that is held in the oceans. Even more recently, scientists have tracked the flow of plastics from rivers into estuaries and then into the high seas, where oceanic currents and gyres (vast vortexes that act almost like the plug hole of a bath) keep plastic waste in perpetual circulation. Battered by the wind and waves, and relentless sunlight, big bits of plastic photodegrade into smaller and smaller bits of plastic. Arguably, this is when they are at their deadliest, as sea life mistakes tiny bits of plastic for food. In storms, the plastic is disgorged on to land and the whole endless cycle seems to start again.

A few years ago, I visited an amazing turtle sanctuary in Watamu, on the Kenyan coast (home to five out of the seven sea turtle species). I got to help carry a huge green turtle back into the water. Having ingested plastic, it was found struggling by fishermen and brought into the Watamu 'turtle hospital', where the staff operated and returned the turtle to health. But as we released it into the waters, I couldn't help but be alarmed at all the plastic washing back on to the land. As the turtle steered off through the foamy white-topped waves, some flip-flops and plastic packaging were rolling in. How long before the turtle ingested more plastic, and next time would it be so lucky?

It's pretty unfortunate that we've decided to use plastic as the primary material for everyday disposable items because it is famously manufactured to last. This means that when you decide, 'I'm going to throw away this bottle/nappy/plastic grocery bag/straw/stirrer/food carton/sauce sachet/sweet wrapper' you're kidding yourself because there isn't really any 'away'. In truth it will take hundreds of years for that plastic to decompose, and meanwhile it degrades into smaller fragments. These are ingested by wildlife (often causing illness or even death) and have been shown to transport disease and toxic chemicals. Plastic chemicals (such as plasticizers, which make plastic ultra-bendy and malleable) have been detected in the human bloodstream. We don't yet know what that effect will be. Landfill dumps are full of plastic, preserved for future generations. I often wonder what those generations (if we are sufficiently lucky to secure enough of a liveable planet to sustain them) will make of this time capsule. They will surely think that we didn't love the Earth very much to pollute so much of it and to keep on manufacturing plastic even when we had 1 billion elephants' worth that we didn't know what to do with.

Volumes of plastic are continuing to increase. They are not slowing down. With increasing volumes adding to the huge mountain we already have, societies are resorting to burning plastic, but as it is made from crude oil, carbon is released when it is burned, and this adds even more greenhouse gas emissions to the atmosphere. In those places where plastic is burned in open pits and there's no infrastructure to deal with the pollutants, inhabitants must breathe in toxic fumes.

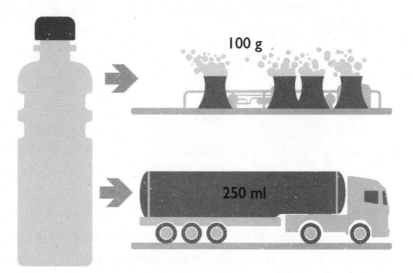

Each 30 g plastic water bottle uses 250 ml of oil
and causes 100 g of carbon to be emitted.

Being a synthetic material, plastic falls under the Planetary Boundary category of 'novel entities' along with 350,000 different types of manufactured chemicals[3] now on the global market. Recently, scientists concluded that this category, together with greenhouse gas emissions, has also crossed a dangerous threshold. We are simply producing too much for planetary systems to deal with. When it comes to plastic, as one scientist who contributed to the research put it, the impact 'includes climate, via fossil-fuel use, land and fresh water systems via use, pollution, physical changes and spread of invasive species, antibiotic resistance genes and pathogenic microbes in the oceans'.[4] These are devastating impacts for Planet Earth and far outweigh any 'good' plastic may do for the environment, such as being an important material in modern cars, helping to make them lighter and thus requiring less fuel.

Did you know?

In the US, according to research by jeans manufacturer Levi's, half of all garments made annually are burned (via incineration plants) or buried (in landfill dumps) within a year.[5] They don't even get to their first birthday.

All this leaves a clear plan of action for us. Of course we're going to need some plastic. It is used, for example, to make replacement heart valves or to enable ocean exploration, but that doesn't mean that everyday disposable items should be made of plastic or that plastic should edge out natural fibres such as wool and cotton in our wardrobes (in a survey of the UK fashion industry for 2020, 60 per cent of the fibres used in online fast fashion ranges was found to be plastic). As a friend of the Earth, it's up to us to spot the ridiculous uses of plastic and to resist the creeping plasticization of life on Earth. I'm a natural dumpster diver and rubbish bin observer. I find it quite difficult to pass a bin without flipping up the lid to see what's inside. I mean, don't you just want to see what's in there?

One of the very favourite pieces of journalism I was part of involved delving into rubbish bins, and was written for a Sunday supplement magazine. I somehow persuaded four families in the UK to save up their rubbish for four weeks. Then I sorted and examined each bit. Along with researchers, I worked out how much was plastic (this is often more complicated than you might think, because plastic is fused together with other materials in food packaging) and then we worked out how much oil was in different components of the packaging; it 'costs' the planet 8 grams of oil for the 'stay clean' lid of a ketchup bottle, in case you were wondering. Once we'd tallied the totals, we presented each family with the results. They were completely shocked, not just by the amount of plastic and oil they had accumulated simply by living their lives, but also by the impact it had and the fact that most of the plastic could not be easily recycled. Even in so-called advanced economies with recycling infrastructure, it is not as easy to reclaim and recycle plastic as many of us think or hope.

This brings me back to my central point: we must fight the pressure to consume plastic, tooth and nail. This is one of the most important things you can do as a rank-and-file 'consumer' in order to uphold your relationship with the planet.

For good measure, for the article mentioned above, we photographed the participants lying in their waste. On the cover of the magazine we featured a baby, Samuel, kicking his babygro-clad feet with joy, oblivious to the rubbish that surrounds him. I often think of that. He is now a young adult, and as for that plastic – well, most of it will still be with us.

When it comes to plastic, many people ask me – in quite a panicky way – what they should do in order to do 'their bit', what easy swaps and little changes they can make to turn the tide on plastic, to which I say: 'Do everything. Do all the swaps, get all the reusables and refillables if you can.' But, in truth, this is less important than understanding the fundamental issue here: plastic is being pushed on us by the fossil-fuel industry and that pressures us to harm the Earth. Of course, we're going to need to use some plastic. Just as we're going to need to use some of those 350,000 chemicals that make up the 'novel entities' Planetary Boundary (although we should surely reduce their impact and make them pollution-free). But we should get together like grown-ups (and I hesitate to use that tired cliché given that kids have shown a lot more initiative when it comes to Earth defending!) and decide what is important, and what is crucial to our existence.

I can tell you categorically that this would not include shrink-wrapped coconut. Nor would it include wet wipes. I first came across coconuts cocooned in an oily, slimy ball of thick plastic when I was working on the household waste story I shared with you above. At the time I had made a big effort to stop shopping at major supermarkets and lived in an area with lots of independent shops, so I hadn't seen them. When I did, I got really angry and fired off a letter to the multinational food retailer in question. I explained that coconuts came in their own hardy packaging, provided by Mother Nature, i.e. a coconut shell, and would they mind explaining why they had decided to 'improve' upon this by adding an extra bit of non-degradable oil-rich plastic that would doubtless be with us for hundreds of years. There were excuses galore. These included the fact that they required the packaging to attach a barcode so that the checkout assistant could recognize the coconut at the till. But my favourite by far was that the coconuts must be wrapped because the fibrous hair provided by Mother Nature represented a choking hazard. In the end, with my army of readers (always so appreciated and so helpful when you're trying to press for change), the retailer conceded that this was indeed environmental madness and agreed to stop wrapping the coconuts.

It felt like a token victory, though, when on a recent visit to a supermarket I noticed a 'drinking coconut' featuring six different types of plastic polymer (none of which can be easily recycled). Not only was it shrink-wrapped but it was sitting on a stand and included a separately wrapped straw. You can imagine my mood!

Similarly, wet wipes also have the ability to cloud my sunny disposition. Just seeing a pack of them can make my eyes twitch and lips begin to curl in a distinctive snarl. The wet wipe is a plastic fibred sheet used to remove make-up and grime, wipe babies' bottoms or even for single-use cleaning. It replaces the flannel, the cloth, the mop and many other sustainable reusable types of equipment that people used to keep themselves and their houses clean for millennia. They are sold as a convenient boon, but in truth they are a scourge on humanity, a blot on the landscape. Turns out wet wipes do not just clean your face, but also appear to be changing the foreshore: the very definition of mission creep.

I live in Greater London near the River Thames, the great watery artery of the UK's capital city. It moves such a volume of water through its estuary and tidal and non-tidal system that it actually qualifies London as a coastal city. In 2021 volunteer litter pickers working at low tide counted 27,000 congealed wet wipes along a 200-metre stretch of the River Thames.[6] In this river, and many others worldwide, congealed wet wipes are changing the shape of the riverbed and impacting on its ecologies in ways no one had begun to anticipate. Beneath the ground they collect in sewers, bringing together all the other stuff you might expect to find in sewers (yes, the waste that we try not to think about). I have had the dubious honour of being part of the first British film crew to come face to face with the UK's largest 'fatberg', as the resultant gargantuan lumps of waste are called. It was every bit as glamorous as it sounds.

Given all the waste we've talked about in this chapter, you may find yourself starting to feel a bit panicky, as if everything is getting out of control. Good. That's exactly the feeling we need to have, because the situation *is* out of control, and it has to end.

Now for the good news. Plastic pollution is increasingly visible – unfortunately that is literally true because everywhere you go you will see it. But increasingly also, there are civil society organizations across the world that are coming together not just to clean up the plastic that is already there but also to say loudly and clearly that we will not add to it (we'll learn more about such organizations later in the book).

When it comes to waste, it is clear that the best way to be a friend to the planet is to produce as little waste as possible. There are some incredible examples of people reducing their waste footprint. But it's also about recognizing when products and behaviours are unacceptable. It's about telling retailers that you will not accept their shrink-wrapped coconuts and demanding that they listen. There is no more shuffling silently away. Before we galvanize ourselves into action, though, let's find out what else we know or don't know about waste … and yes, that includes quite a few questions on plastic pollution.

CAN YOU TURN RUBBISH TO TREASURE?

YOUR ESSENTIAL TEN-QUESTION QUIZ

1. In 1987 a so-called 'garbage barge' hauling 32 tonnes of waste became an unlikely star of US nightly news shows as viewers followed its travails over many months. The barge, containing household trash from New York, was owned by Lowell Harrelson, a waste entrepreneur who had decided to try to avoid New York's main landfill as it was close to full and charging high fees to dump. He came up with the bright idea of steering the barge to Central America and dumping the trash there. In common with many 'bright ideas' this turned out to be more difficult than expected; none would admit the trash, leaving the barge homeless. But what was the name of the legendary NY landfill site that Harrelson was trying to avoid?

A. Liberty Landfill
B. The Staten Pits
C. Fresh Kills
D. Rogers Salvage and Scrap

2. What is the name of the huge ocean rubbish patch that currently floats between Hawaii and California covering an estimated 1.6 million square kilometres?[7]

    A. The Great Pacific Garbage Patch
    B. The Pacific Vortex
    C. The Five Gyre Rotator
    D. The Great Plastic Soup

3. When it comes to stuff, we need to make better decisions about materials and to wise up to waste. So for this question, you need to select the correct pair. What is the most used plastic resin also discarded as waste and what is the biggest source of litter (fugitive waste) on the planet?

    A. Polyethylene and cigarette butts
    B. Polystyrene and drinking straws
    C. Polyethylene terephthalate and plastic water bottles
    D. PVC and single-use face masks

4. How many plastic beverage bottles do you think are bought each minute?

    A. 120 billion
    B. 1 million
    C. 200,000 million
    D. It doesn't matter because most are collected and recycled back into bottles

5. The plastic products we consume on a daily basis are manufactured all over the world. To feed these engines of plastic production, tiny pre-production pellets of plastic are transported globally by cargo ships and trucks across the world. Millions of tonnes escape, an astonishing 230,000 tonnes of these into oceans each year, causing extraordinary damage to ecosystems and organisms. In the plastic industry they are often referred to as 'nurdles'. But what is the undeservedly poetic name that has also been given to these eco-pests?

A. Twinkles
B. Sea jewels
C. Mermaid's tears
D. Industrial coral

6. We know that the Earth is infested with tiny threads of plastic from the increased use of synthetic materials for clothes and fashion. These are under 1 millimetre thick and are known as microfibres. When we wash garments, thousands of them are released into our waterways. But which type of garment is the biggest culprit?

A. Christmas jumpers
B. Football/soccer shirts
C. Fleece jackets
D. Waterproof anoraks

7. In the early 1970s, 2 million tyres were dumped on the Atlantic Ocean floor off the coast of Florida, for what reason?

A. A conservation group decided to build an artificial reef
B. Criminal gangs were pretending to recycle
C. It was hoped they could stop coastal erosion
D. A major studio needed them for an underwater sequence in a James Bond film

8. The Indonesian biologist Prigi Arisandi has become well known in his country as a waste warrior, travelling across Indonesia with a team of volunteers in hazmat suits to remove which polluting items from waterways?

A. Face masks
B. Plastic drinking bottles
C. Nappies
D. Footballs

9. Despite the UN in 2015 calling for an end to open defecation, nearly 950 million people still have no access to toilets. A safe place for number ones and number twos remains number six on the UN's list of Sustainable Development goals. Which politician campaigned with the slogan 'toilets before temples'?

A. Narendra Modi
B. Mahatma Gandhi
C. Benazir Bhutto
D. Judy Wakhungu

10. For our final question on waste, let's head to another part of the solar system (yes I know this book is all about the Earth, but human litter is causing an issue literally outside this planet's boundaries). Specifically, the US Department of Defense's global Space Surveillance Network (SSN) sensors currently track around 27,000 chunks of human-made space junk travelling at high speed in the near-Earth space environment[8] In 2006, astronaut Piers Sellers was carrying out an experiment when he accidentally 'littered' which object?

A. A spatula
B. A test-tube
C. A toolbox
D. A spirit level

ANSWERS

1. The answer is C, Fresh Kills. The famous landfill site on Staten Island was once so big it was rumoured it was visible from space, like the Great Wall of China. The dump opened in 1947 and the name 'Kills' is derived from the Dutch word for 'steam'. For half a century it dealt with immense quantities of rubbish from New York – one of the most wasteful cities in the world. By the 1980s it was filling up and by 2001 it was closed to municipal waste. The incredible 3.2 million tonnes of trash produced by New York on an annual basis today is dealt with well beyond city limits – most reprocessing and incineration takes place in South Carolina and Pennsylvania.[9] Following 9/11, Fresh Kills became a burial site for victims of the World Trade Center attack.[10] As for the garbage barge, it eventually rumbled its way back to New York, followed by TV cameras, where Mr Harrelson finally coughed up for the landfill fees. However, the whole sorry saga and the spectacle of household rubbish on the 'garbage barge' is also credited with kick-starting New York's recycling revolution.

2. The answer is A, the Great Pacific Garbage Patch, or the GPGP. It isn't the only plastic gyre – five rotating gyres pull plastic waste into a trash vortex in the Earth's oceans. But it is believed to be the biggest and is definitely the most famous build-up of plastic. An important scientific study (published in March 2018) found that the GPGP was sixteen times bigger than previously thought. Stretching across 600,000 square miles of ocean, it dwarfs France, is bigger than Texas, weighs in at 79,000 tonnes and contains an estimated 1.8 trillion pieces of rubbish, 99.9 per cent of which is plastic.[11] One item pulled from the patch was found to be forty years old – talk about legacy plastic!

    If you went for D, it may be that you've heard the GPGP described somewhere as a plastic soup. Those who have sailed through it report that rather than it consisting of endless big chunks of recognizable plastic, it contains pieces of various sizes and density, giving it more of a soup-like consistency. Please, let's not reduce mighty and majestic oceans to sad rubbish soups.

3. The answer is A, polyethylene and cigarette butts. Plastic production has grown exponentially since the 1950s. Almost all of the plastic produced still exists, held in the ocean as pollution, in landfills and in some durable products. Just 9 per cent of plastics are recycled globally each year as plastic production continues to grow. Plastic 'litter' surveys are taken primarily from beach and land cleans. Four and a half trillion cigarette butts – which contain plastic and are non-biodegradable – are estimated to be discarded as litter each year.

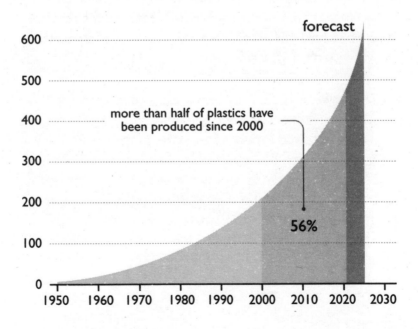

Global plastic production continues to rise.

4. The correct answer here is B, 1 million.[12] That's 20,000 a second or 480 billion a year.[13] That's enough to extend more than halfway to the sun if the bottles were placed end to end. If you said A, 120 billion, you might have been thinking of the global total of plastic bottles produced by just one soft drinks brand (yes, it's Coca-Cola[14]). This total drew gasps of disbelief when it was calculated and revealed by NGO Greenpeace a few years back. If you answered D, well you really are one of life's optimists. Only 9 per cent of plastic is recycled globally. I'm ending this answer with a plea: if you live in a country with a secure water source (i.e. where potable treated safe certified water is available twenty-four hours a day from a tap or faucet) then please stop buying bottled water.

5. It's C, mermaid's tears. In truth, I imagine the mermaids are bawling their eyes out. Mermaid's tears, aka nurdles, are thought to be the second-largest source of micropollutants in the ocean, by weight, after tyre dust.[15] In May 2021, the MV *X-Press Pearl* sank off the coast of Sri Lanka in what was called 'the worst catastrophe in [Sri Lanka's] maritime history'.[16] Eighty-seven containers full of lentil-sized mermaid's tears, adding up to 1,680 tonnes of nurdles, were accidentally disgorged during this disaster. In places around Negombo, post-disaster conservationists reported the mermaid's tears on the coastline to be 2 metres deep. Many called for these so-called mermaid's tears to be reclassified officially as hazardous waste and made the point that while bunker fuel from ships and crude oil tends to be the main area of concern when ships run aground, tiny pieces of microplastics are perhaps the most serious threat to the oceans. One of the major issues with pellets of plastic in water is that they are often mistaken for food by wildlife. Research shows that any creature consuming them feels full afterwards, only to die of starvation through having a stomach stuffed with plastic.

6. It's C, fleece jackets. During a laundry cycle, a fleece jacket can 'shed' over 250,000 microfibres.[17] Anything made from synthetics can shed tiny fragments of plastic, but because of the loose strands that characterize fleece garments, they are more susceptible to shedding. If you use a top-loading laundry machine, then the effects are even worse (as this causes extra agitation in the wash cycle). The textiles and garment industry is working on ways of creating synthetic fibres that stay rooted during the wash. But it's not clear how these will work. We are also starting to see washing machine filters that can be fitted to catch microfibres. But for now, I prefer to shun fleece and as many synthetic fibres as possible.

7. The answer is A, the tyres were dropped into the Florida waters to form the Osborne Reef. That might be unexpected; after all we don't associate conservation groups with the dumping of millions of tyres these days! But incredibly, this was seen as a win-win situation; tyre recycling was in its infancy and this seemed to offer a solution, while, at the same time, conservationists were exploring ways to rebuild the reef. This tale represents a lesson in the evolution of Earth-saving – we don't always get it right. What's important is that we move to rectify mistakes as we learn more. In the case of the Osborne Reef, ocean groups have been monitoring the tyres for some time. Because the area in which the 'reef' is situated is increasingly being hit by tropical storms (not something the 1970s crew could necessarily have foreseen), the tyres have broken free and are actually damaging the natural coral as they are dragged along the ocean floor. Enter global groups such as 4oceans, which are physically rectifying the mistake and removing as many tyres as possible. This is likely to be a job that goes on for a long time, so if you find yourself in the Florida area and you have some diving skills, look the group up and lend your help.

8. The answer is C, nappies. Prigi Arisandi, the recipient of the prestigious Goldman Environmental Prize, leads the Diaper Evacuation Brigade attempting to fish used nappies out of Indonesian rivers. According to an interview Prigi gave to the *Guardian*, his epiphany came when he stood in the Surabaya River in East Java and counted 176 nappies floating past in one hour. Indonesia produces an estimated 6 billion disposable nappies every year. A 2018 World Bank Study found that disposable nappies made up 21 per cent of the waste in water in fifteen major Indonesia rivers. Yuk.

    Diaper or nappy waste is now a major source of pollution all over the world; in the UK an estimated 8 million nappies are sent for landfill or incineration every week. Such eye-popping figures have prompted many new parents around the world to seek out eco-alternatives, including 'nappy libraries' – the renting of sets of reusable/washable nappies. This is an example of the planet-approved sharing economy, which we'll talk about in greater detail in the next chapter.

9. The answer is A, Narendra Modi, the premier of India. If you thought it was Mahatma Gandhi, he did say 'sanitation is more important than independence'. But still today poor sanitation and unsafe water kill more children than measles, malaria and AIDS together: some 1.4 million. Living better together (a concept the Earth approves of!) isn't of course just about buying better stuff or from different shops (although that does play a role, as we will find out in the next set of questions) or just thinking about ourselves. It means we also need to lend our voices to support global organizations that dedicate every minute every day to making change a reality.

    Today, we are lucky that we have defined targets in the form of 17 Global Goals, commitments made by 193 world leaders to end extreme poverty, inequality and climate change by 2030. Wherever possible, lend your support to these goals and the companies and organizations that support them. I'd recommend finding out more from organizations such as Water Aid Global (wateraid.org) and seeing how you can help.

10. The answer is A, a spatula. Yes, Piers Sellers was testing heat tile putty using a spatula when it flew off into orbit, adding to the thousands of pieces of space junk or debris already out there, which range from random objects to defunct satellites. The volume of space junk concerns organizations such as NASA because if it becomes too great it could jeopardize space exploration for future generations. This real threat has provided a plot line for a famous movie; in the film *Gravity*, Sandra Bullock and George Clooney's attempts to fix the Hubble telescope are thwarted by space junk.

At least the Sellers Spatula was spotted re-entering the atmosphere four months after it flew off, dropping as ash into the Atlantic Ocean. So it's no longer an active high-speed threat in space, but we have to hope it didn't add any microplastics from the ash when it entered the ocean. Perhaps it's a reminder that everyday items can end up in some extraordinary situations.

So, how did you do? I'm willing to bet that you did quite well as your antennae for spotting the Earth-friendly answers starts to develop. But if not, don't worry. There were a few curveballs in there and a few strange names. You'll definitely remember them next time. But whatever your score, I'll also bet there is work to be done in making your own trash footprint smaller. Because yes, it's important that we overhaul waste systems, pressure our governments to stop dumping rubbish elsewhere and lobby for a global treaty to stop plastic waste, but a really important way of demonstrating this is to start with your own consumption.

8

LIVE YOUR BEST (EARTH-FRIENDLY) LIFE

We've trawled through our rubbish bins and taken a forensic look at our kitchen cupboards, all with the planet uppermost in our mind. But we are not the only ones doing this, I promise you. Across the globe millions of people in all walks of life are responding to the nature and climate emergency. We'll meet a tiny proportion in Chapter 10 when we'll test your knowledge of heroic climate and nature action. But so many people are also putting the Earth first in their professional lives, whatever they do.

There are crazy, mountain-like challenges we need to rise to in order to restore the Earth to tip-top health, but there are also spectacular opportunities. Those of us in high-consuming countries with our big-footed lifestyles need, of course, to scale down our consumption, but entrepreneurs, innovators and disruptors have a golden opportunity to help us reinvent the way we live, relax and consume, and to make each of these better as well as greener (in my experience the two often go together). Much of the way we live and consume now could be improved. We spend huge chunks of our lives sitting in traffic jams, we breathe in horrible dirty air that makes us ill, we end up swimming in sewage, and we get stressed by things like credit card debt after buying things we don't always need. Lots of research shows that seeing plastic pollution actually causes us psychological damage. We feel it very emotionally.

To me it makes so much sense to talk directly to the people who design our lives – from town planners to handbag designers, and ask them if they can prioritize the Earth and just make things better. I have always loved to talk to designers. I don't care what they design – clothes, widgets, jam-jar lids – they always have a solution to hand. I don't believe anybody goes to college and learns their skill with the intention of designing for landfill. In secret, they all dream of coming up with an epoch-changing solution that will rescue humankind, end deforestation and be talked about until the actual end of time (which will be further off than 2050 thanks to their help).

This is the type of glory hunting we need to play on. The trouble is that some everyday things we use were designed so long ago that they aren't at all helpful. Almost all of the problems we currently have are centred on the amount of resources we extract from the Earth – and the waste that we cause can be attributed to our lazy old linear economy. 'Linear' is exactly how you'd imagine it – a straight line of actions that *steal* resources (I know that's an emotive word, but it's how the planet currently feels) to make stuff that is then pushed out into the world without any plan for how we will get those resources back in order to reutilize them. The linear economy is very good at producing stuff that we don't much care about, but it is a very weak friend to the Earth. Don't you think we could do better?

Did you know?

Around the world, people are using their smartphones for increasing periods of time before upgrading them, which makes them more sustainable. In 2019, users in five European countries – France, Germany, Great Britain, Italy and Spain – were found to have extended the lifecycle of their smartphone by nearly three months, from 23.4 to 26.2.[1]

But how? Well the answer to replacing a straight line is to pick a different shape. And that shape is a circle. Circles and the planet go together like the proverbial horse and carriage. After all, nature is all about circles. The Circle of Life is not just a song sung by animated lions, it is the basis of biological and geological life. So when we're looking for an eleventh-hour solution for how to fix our planet, it kind of makes sense to start with a circle.

Circular-economy thinking sounds quite shiny and out of reach, but it's actually very simple. We redesign everything so that materials stay in use and are constantly redeployed. In short, we're designing out waste. Some have translated this as continuing to push out huge amounts of plastic tat and then trying to come up with ways of getting it back and recycling it. You'll often see the 'fast' industries claiming that they are bringing in all manner of circular recycling systems. These are often advertised through 'takeback schemes'. For example, a fast-fashion brand will tell you to drop your clothes back instore when you don't love them any more in return for a voucher that will incentivize you to buy some more (explain that one to me!). Or a toy company will run a takeback scheme for old toys. These schemes give the impression that old clothes and toys can become new clothes and toys. It is not that simple.

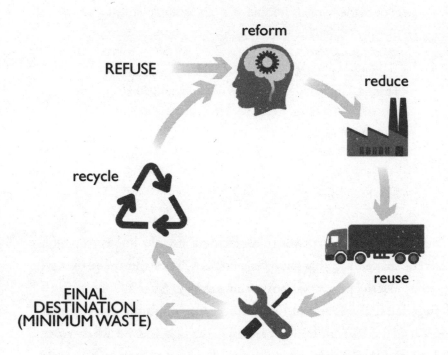

Designing out waste through circular recycling systems.

Most consumer goods now contain plastic and it is complex to recycle that into the same format. So, a garment might become filler for car seats and a toy might become part of a park bench. This is downcycling, where the material loses value each time it is reclaimed and is unlikely to be recycled more than once. So, not a circle. To make a circle, you need to plan from the outset – and it doesn't necessarily need to involve recycling after one owner. We need to think bigger. MUD Jeans, for example, is a brand that produces denim jeans (as the name suggests). Once they produce a pair of jeans, the focus is on keeping that fabric going around the circle for as long as possible. This is because you don't own MUD jeans, you lease them. When you've done with them, they are returned and either mended, washed and leased again, or remade into new jeans. To get the circular economy rolling forwards, we need more leasing schemes like this, as well as more materials that are genuinely easy to recycle, such as aluminium, and we need more sharing.

Did you know?

One in every four bananas sold in the UK is now Fairtrade (a scheme through which developing-world producers are paid a fair price for growing). That's a big deal in a country where 5 million bananas are consumed each year.[2]

Yes, sharing is caring when it comes to the planet, and this is one of my favourite slogans. There are so many things that clutter up our lives and that we wouldn't even notice if we didn't actually own them. The power drill, for example, is famous for its lazy lifestyle. It sits in your toolbox for a lifetime but is used for an average of thirteen minutes. Yet so many households own one. This is crazy. Every street or block could have a tool library, housing these power tools and loaning them out. Less clutter for individual households and a whole new business model.

One thing is certain – if we grow crops that help the planet, such as nitrogen fixers and nutrient replenishers, we will go some way to help the Earth regain its mojo. If we are able to move from a bad way of making stuff – i.e. from plastic-based textiles for fashion – to a good way, using regenerative textiles (such as those made from crops) to make clothing that sequesters carbon and produces a healthy ecosystem, then we will take huge strides in becoming the ultimate friend of the Earth.

SHARING IS CARING

If you are a designer or maker, look out for resources that are open access (where resources and knowledge are given away free online). These platforms and ways of sharing knowledge and innovation are the tools you need to be a better friend to the planet. Materiom, for example, is a platform founded by circular-economy specialists and gives open data on materials made from renewable resources. This is a great treasure trove for product designers who are determined to be circular, not linear!

Fortomorrow.org is brought to you by the United Nations Development Programme Accelerator Labs, which have expert spotters across the globe, from marketplaces to colleges, looking for innovators and scaling-up solutions. Perhaps you have your own great idea to pitch to them?

In the very near future we will have access to many more products and services that make much more sense to our friend the Earth, and automatically lower our footprint. We might even design or come up with some of these products ourselves. Many global economies have introduced net-zero laws and these will drive progress as we switch from polluting technologies to ones that have a much smaller impact. Once these shifts start to happen, they can accelerate pretty fast – so hold on to your hats!

But for now the best way of keeping on track is simply to prioritize what's best for the Earth in your everyday decisions. It doesn't have to be complicated. You don't need to spend hours choosing the most ethical type of loo roll. But you can teach yourself to recognize the danger points, and the appropriate badges and certification (such as the Forestry Stewardship Council (FSC) logo, which tells you a forestry product comes from a sustainably managed supply chain), to make sure your hard-earned cash is going in the right direction. After all, you don't want to be funding anything that is destructive to your dear friend, Earth.

I don't really like the term 'consumer' because I think we are all global citizens with agency, and this belittles us and makes us feel as though all we can do is buy stuff from Jeff Bezos and hope for the best. The term 'ethical consumer' is more helpful – it still indicates consuming (as we all do to some degree) but means making better decisions that respect the welfare of your fellow humans, flora, fauna and the biosphere – everything we've talked about so far basically. Thanks to movements such as the circular economy, it should become easier to be ethical consumers by consuming very little. Rather than agonizing over whether or not something fits a whole range of criteria and is more or less bad than another product, the one known and proven way to take the sting out of any consumer product is to extend its lifespan for as long as possible. Take clothing, for example. Instead of spending hours debating whether the Earth will approve of your new sweatshirt made from 100 per cent organic cotton, which is actually a more complicated question than you might think, wear that sweatshirt every day until it falls apart. That's the way to make it Earth-friendly. Rewearing, reusing, remanufacturing, refashioning, repairing – these are the new noble arts. The more of this you do, the more the Earth will thank you.

OK, so let's get ready for ten pollution-busting questions covering footprints, repairs, useful plastic and billionaire lifestyles, and everything in between.

WINNER TAKES ALL
OR FAIR SHARES?
YOUR ESSENTIAL TEN-QUESTION QUIZ

1. There are huge, and I mean MASSIVE, differences in the amount of emissions households are responsible for, depending on where you happen to be born and live. It's no surprise that it's those of us in developed, consumerist economies who are hoovering up very big slices of the carbon pie, leaving little for anyone else. According to research by Berlin's Hot and Cool Foundation, what do we need to drop our individual carbon footprints to by 2030?

A. We need to get to 2.5 tonnes per year per capita

B. We must aim for 1.4 tonnes per year per capita

C. We must get down to 0.7 tonnes per year per capita

D. We must get our carbon footprint to zero immediately

2. The richest 1 per cent of all the people on Earth are responsible for more than twice the amount of carbon pollution (greenhouse gas emissions) as:

A. The whole of Bangladesh

B. The poorest half of humanity

C. The United Arab Emirates (UAE)

D. Monaco

3. Earth Overshoot Day is the annual marker of the date when humanity's demand for ecological resources and services in a year exceeds what Earth can regenerate in that year. It is calculated by international research organization Global Footprint Network. What is the call-to-action hashtag that has been used in recent years?

A. #EatLessMeat

B. #BuyLessStuff

C. #MovetheDate

D. #LifeBalance

4. In order to do better, we also need to avoid being led up the garden path. Increasingly, legislation is being developed to stop companies and authorities from over-claiming on their green efforts or selling goods on the basis of sustainability when the claims aren't verifiable. What is this form of deceit called?

A. Greenwashing

B. Woke-ing

C. Jerrymandering

D. Eco-inflating

5. The Fairtrade movement exists to give producers in developing countries a fairer foothold in international trade and includes environmental safeguards. Over the years, the Fairtrade movement has blossomed, and you can find thousands of products bearing the Fairtrade Foundation mark on sale all over the world. But what was the reaction from supermarket bosses in the UK when they were first pitched the idea of selling Fairtrade on their shelves?

A. They doubled their orders within two weeks
B. They said only vicars would buy it
C. They began by trialling Brazil nuts
D. They put the Fairtrade products at the front of the store

6. As I've stated earlier, sometimes plastic is very useful and creates products that help rather than hinder. This is a point reinforced by the Museum of Design in Plastics (MODiP), located at the University of Bournemouth in the UK. Object number AIBDC: 005910, held on permanent exhibition, was invented by one Wendy Brodie of High Wycombe in the United Kingdom.[3] But what does it do?

A. Forms a plastic skirt that catches plastic in waterways
B. Remanufactures empty plastic bottles at home
C. Acts as a human squeegee to displace laundry
D. Attaches to the crisperator unit in the base of the refrigerator to extend the lifespan of fruit and veg

7. In 2009 two US graduates Jennifer Hyman and Jenny Fleiss set up an e-commerce platform that attracted a huge buzz, millions of dollars of investment and promised to disrupt a notoriously eco-unfriendly industry. But how was their disruptive business model described?

A. Netflix for Dresses
B. The Davos of Downloads
C. Tesla for smartphones
D. Amazon for eco-cosmetics

8. You might not have realized it but we're in the midst of a global revolution, led by the Right to Repair movement. In some countries (including the UK, France and the US) legislators have begun to bring in so-called right to repair laws, giving 'consumers' the right to repair and access spare parts for all kinds of consumer items, from refrigerators to washer dryers and TV sets. But what is the Earth-hostile phenomenon that this movement and these laws seek to stop?

A. Moore's law
B. Occam's razor
C. Overton's window
D. Planned obsolescence

9. Before we sign off in a blaze of glory, let's focus on what *not* to do. I'm really hoping that you're in this for the long haul and that your commitment to Earth's wealth and health will continue even if your circumstances change. But if you do happen to join the world's elite 1 per cent, the super-rich, or you are a member already, which of the following should you definitely avoid because it will engorge an already super-large carbon footprint?

A. Travelling in a private jet
B. Owning more than four houses
C. Possessing a superyacht
D. Owning racehorses

10. Now here's your chance to employ many of the skills you've learned so far. We know living better involves making smarter decisions, and that can mean choosing those with the lowest impact. As we've discovered, every product has a secret backpack of resources and an energy footprint from the energy used to manufacture it. Those calculations give us a product's break-even point – the point at which the number of uses or length of time that product is used for surpasses the energy that went into it. To live better, we need to be able to pick the product that will last longest, working off all the

stuff that went into it in the first place. But which two of these break-even points are real and which are complete fiction?

A. A reusable cup used 21 to 27 times will break even (compared to a single-use paper cup)

B. A hairdryer needs to be used for 20 minutes a day for one year to reach break-even point

C. The break-even point of an electric car is nine years

D. The break-even point of a *faux* Christmas tree is a decade

. . . .

ANSWERS

1. The answer is A, 2.5 tonnes by 2030. This is not going to be easy for someone like me in the UK where carbon footprints weigh in at 10 tonnes per person per year, double the global average. But the work doesn't stop there. Oh no! The Hot or Cool Institute researchers found that once we've got to 2.5 tonnes per year, then we must go further; by 2040 it needs to be at 1.4 tonnes and by 2050 we need to be living very lightly at just 0.7 tonnes per person.[4] If this sounds impossible, well, nothing is impossible, and we do have to rely on a huge worldwide transition that will make increasing amounts of our infrastructure 'green'. Phew, that means we don't have to take responsibility for everything ourselves. But it's also important to note that many people need to *increase* their carbon budgets to improve their standard of living (more on this in the next answer). In order for them to live sustainably, though, in the future, they too will need to shift to Earth-friendly technologies, especially renewable energy, as soon as possible. In short, we need to balance this out.

2. Elaborating on the previous question, the answer is B, the poorest half of humanity,[5] which numbers 3.1 billion people.

As we know, in order to help out the Earth (and ourselves) we must slash our carbon emissions, and that means living on a reduced global carbon budget. But as developing and emerging countries can argue with some justification, when the budget is split so unfairly why should they scale back at the same rate as developed nations? The youth climate activist Greta Thunberg wrote in a newspaper editorial in 2019 that 'The bigger your carbon footprint, the bigger your moral duty',[6] and many agree with her (including me!).

3. The correct answer is C: the challenge for us globally on Earth Overshoot Day is to #MovetheDate. The day of the year on which humans have exhausted the Earth's regeneration capacity has been creeping up ever earlier in the calendar annually over decades. In 1970 Earth Overshoot Day fell right at the end of the year, on 30 December, whereas in 2021 it fell on 29 July. Yikes! Anything we can do to help push that date back towards 30 December needs to be done. Ultimately, of course, we want to retire the date altogether! Make sure you put a note on your family calendar to check overshootday.org each year. Just make it a different month to when you file your tax return in case it all becomes a bit too depressing!

4. The answer is A, greenwashing. Its origins are actually in the travel and tourism industry. In the 1980s an American environmentalist, Jay Westervelt, who was evidently doing a lot of travelling, became incensed by the way hotels put signs up pleading with guests to reuse their towels, thus 'saving the environment'. He figured that they were doing little to promote recycling elsewhere and he suspected the primary motive of most establishments was to save on laundry costs.

5. The answer is B, I'm afraid. Pioneers of the movement hawked Fairtrade tea and coffee samples around supermarkets to be greeted with the response, 'Only vicars would be mad enough to buy those', meaning only vicars or

other people of the cloth would be pious enough to care. However, the gatekeepers to the nation's shopping trolleys begrudgingly gave it a try. And they were wrong. The general public loved products offering a premium that would go back to the producers. From the initial tea and coffee, there are now over 1,500 Fairtrade retail products for sale across the world. These include Brazil nuts, which brings us to answer C, a decoy. The first ethical investment fund was nicknamed 'Brazil' by traders, on account of the fact that you'd have to be nuts to invest in it. Again, ethical investments have become very popular, so they were proved wrong too. It turns out it's cooler to care than people often think.

6. The answer is C: a plastic, injection-moulded device registered as the Bodyflix, reminiscent of the horsehair plait the Romans apparently favoured at the baths! I have one for my household that I bought about ten years ago, but sadly it is now hard to find them on sale. This is a shame because it is what we call a displacement device. Towels are the gas-guzzling SUVs of bathrooms. We wash them far too often at too high a temperature, and many households add on the extra burden of using a mechanical dryer to dry them. (If you wash your towels at 30 degrees rather than a higher temperature, you use around 40 per cent less energy). So, instead of relying on heavy towel use with a big laundry footprint, the human squeegee enables you to scrape off water in the shower cubicle and then you can pat dry with a towel.

7. The answer is A, the Netflix for Dresses.[7] The two Jennifers, or Jenny and Jenn, took the fashion and tech worlds by storm with their platform Rent the Runway. The idea was simple — to take high-price designer dresses and share the cost by renting them out to subscribers. But many were quick to seize on the positive environmental implications that this model, based on a sharing economy rather than individual ownership, might have. As we know from elsewhere in this

book, the fashion industry is plagued by disposability and fashion waste. Rent the Runway spawned a new generation of fashion rental platforms, many of which have overt Earth-friendly messaging and seem to be gaining traction with younger fashion consumers. There are also platforms for specific items such as luxury handbags. As well as renting garments to wear, users can list pieces from their own wardrobes to loan. Is this something you can imagine doing?

From the clothes you wear to the miles you drive,
there's always a carbon footprint.

8. The answer is D, planned obsolescence, a strategy that can be traced back to the Great Depression era in the US. The idea – articulated in a pamphlet from the time – was that, in order to stimulate the economy, personal goods should be retired asap and replaced by the manufacture of new ones. OK, so that might have had some currency at the time, but

fast-forward to our era of ecological emergency and you can see that it sounds disastrous. Designing in short lifespans to consumer electronics and white goods gives the Earth a terrible headache. How intentional this is on behalf of brands and manufacturers remains a hotly debated topic. But an example of planned obsolescence that we often come across would be a part with a battery – which has a finite lifespan – that cannot be accessed, perhaps because it's moulded into the product. This means we mere mortals cannot 'fix' the product and instead have to chuck it out and replace it with a shiny new version. Key to stopping planned obsolescence (and the toxic waste it generates) is aligning our rights with the Earth's rights. If you want some extra motivation, head to the tear-down and repair website iFixit,[8] where the mantra is 'we have the right to open, tinker with and repair everything we own'! Spoken as a true friend of the planet. If you plumped for A, that's understandable. Moore's law decrees that the computing power bought for a certain amount of money doubles every eighteen months and is closely connected to obsolescence. I'm afraid you don't get the point, though.

9. The answer is C: possessing a superyacht. We now have a greater understanding of the carbon cost of the billionaire lifestyle, thanks to a study of the same name. Researchers combed eighty-two databases of public records and determined that the superyacht is the great billionaire gas guzzler and climate botherer. The study concluded: 'a superyacht with a permanent crew, helicopter pad, submarines and pools emits about 7,020 tons of CO_2 a year, according to our calculations, making it by far the worst asset to own from an environmental standpoint'. That meant that the billionaire Roman Abramovich, known for frequent sojourns on his yacht, the *Eclipse* – the second-biggest yacht in the world at over 162 metres – topped the list (nothing to be proud of).

10. The answer, my friends, is A and D. If you picked A and D, you are a WINNER! According to studies that compare different materials, you will need to use a ceramic coffee cup 21 to 27 times to get the better of paper.[9] This should be easily achievable; a cup is for life, not just for breakfast! An older, very influential study from 1994 by Professor Martin B. Hocking of the University of Victoria in Canada suggests that paper is much more efficient than this, and that ceramic cups must be used over many years to outpace the energy used to create them and to keep them clean.

    But things are different now; most paper drinks cups are lined with plastic and are complicated to recycle. In the past we have overestimated how many of them get recycled. When it comes to Christmas trees, most fake or faux trees are made from plastic and most are imported from China, which does a roaring trade in Christmas paraphernalia. But if you have an efficient factory versus a family that drives many miles in a gas-guzzling SUV in order to collect their real tree, then the real tree will lose points in terms of its emissions. This is complicated, but a good study, by academic Dr John Kazer, of the virtues of real versus plastic in the UK, where 7 million real trees are sold each year, puts the carbon footprint of a 2-metre artificial tree at around 40 kilograms, which is twice that of a real tree that ends its life in landfill, and more than ten times that of a real tree that is burned.[10] So Dr Kazer argues that you would need to reuse your tree for ten Christmases – at least – to keep its impact lower than that of a real tree.[11]

Old habits can be hard to give up, but I hope you're feeling as if you are learning some new tricks. If you scored low in this chapter (3 or under) you probably still need to let go of some of those old aspirations. Perhaps you are secretly hankering for a vacay on a superyacht or thinking that disposability equals convenience. In which case, think about how amazing you're going to feel when you put that behind you and focus on your connection with the planet, something infinitely more satisfying and fulfilling (and more likely to happen). If you scored anywhere from 4 to 7, well done – you are building good Earth-acumen and could do really well in these final two rounds. If you scored 7 to 10 then you're on course for a spectacular finish. On that note, are you ready for the home stretch?

HOW TO BRUNCH WITH THE PLANET

As we've got to this point in our mission, I'm celebrating with a hypothetical brunch with the planet (my shout). But there's an issue. Actually, there are many. The venue is light and airy, but I'm getting hot and flustered. The heat is caused by embarrassment, because it's incredibly awkward to order food in front of the Earth. Everything on the menu seems so loaded with ecological insult. 'I'll have the waffles …' No, wait – what if they contain palm oil linked to the destruction of the Sumatran rainforest and the collapse of orangutan populations through habitat loss? Why does everything seem to involve a poached egg? Is contemporary egg production compatible with planetary boundaries? And as for meat, well, isn't that driving greenhouse gas emissions? The Earth looks straight at me. 'Never mind that,' she says, removing the menu from my hands. 'What is going on with food waste?'

Did you know?

Globally, one third of all food ends up as waste.[1] In UK restaurants, the equivalent of one in six of the 8 billion meals served by the food service industry every year is effectively scooped straight into the bin.[2]

The fact is that one quarter of all carbon emissions can be attributed to wasted and lost food. (We tend to refer to it as 'lost' when it is wasted in production and manufacturing, so at the farm and factory. We refer to it as 'waste' when it is dumped by consumers. Both are huge drivers of food waste.) According to a recent study published in *Nature Foods*, it's not just that the global food system is affected by greenhouse emissions; food itself is *driving* the climate crisis, and is responsible for 35 per cent of all global greenhouse gas emissions.[3] I hardly need to tell you that climate change is not great for growing food. On the frontline of both food production and the climate crisis, farmers are grappling with increasing climate risks to their livelihood and to the global food supply. As temperatures rise, crops such as coffee and bananas become more difficult to grow.

This reality is starting to dawn on privileged cosseted consumers too. Last summer in the town where I live, there was a mild but audible panic as shelves ran dry of pasta. This was shocking, particularly for those with a limited recipe repertoire and small children. I didn't see any actual tears, but I could tell some were close to it. What on earth was going on? Well, it transpired that this fusilli shortage was a direct result of the summer's very high temperatures thousands of miles away near the Canadian wheat-growing belt. This had then been followed by an unprecedented drought and extreme weather patterns in Europe. All of these factors, influenced by climate change, came together to clobber the global durum wheat crop.

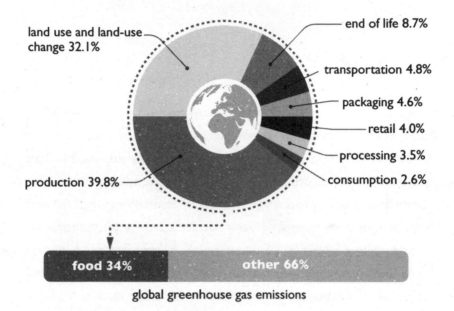

land use and land-use change 32.1%

end of life 8.7%

transportation 4.8%

packaging 4.6%

retail 4.0%

processing 3.5%

consumption 2.6%

production 39.8%

food 34%   other 66%

global greenhouse gas emissions

Slicing up the food-system pie.

For many years, pasta brands in the supply chain have been researching into wheat – trying to find better, more resilient strains, for example. This time they were outpaced by climate change. For some 'consumers' (and you know I don't love that word) this will have been the first time they had even an inkling that something was wrong. In many societies our food shopping is so divorced from the way food is produced that we might as well push our grocery cars around Disneyland. In the UK our supermarkets can carry 65,000 to 70,000 different lines of products, mainly food. We assume full shelves equals food security. That is not the case.

Did you know?

Avocados are seriously thirsty! It takes 227 litres of water (60 gallons) to grow a single avocado according to the Water Footprint Network. So that you have something to compare that to, it takes 240 litres to produce a mobile phone.

We can feel divorced from the growing and production of our food, not least because in many economies it is imported from thousands of miles away, and heavily processed. Our food cultures today are incredibly different from the attitudes of our grandparents. We eat much more processed and out-of-season food, which requires huge amounts of energy to grow under glass or polytunnels and transport long distances. This energy is largely provided by fossil fuels, and that makes our diets heavy carbon emitters. As a result, eating as if the planet matters might not come that naturally to many, but people can be convinced. And the reason they can be convinced is that there are some big benefits for them too. Earth-friendly food means fresher, more local food, a more diverse and considered diet. It also means a healthier diet, and one that tastes better, which, after all, is the point. Getting started on a more Earth-friendly diet needn't be that complicated either. A few firm food rules will point you and your shopping trolley in the right direction.

First, introduce more plants! We seem to struggle to do this, but following standard advice to eat more fruit and vegetables would cut our collective diet impact by a huge 17 per cent,[4] largely because this great habit displaces meat and fish, which have a higher impact. For the highest gains, as many of your five a day as possible should be local.

Second, go seasonal. Eat a wide supply of seasonal food. Think beyond fruit and vegetables. We buy fish as if it was completely divorced from natural breeding cycles – it isn't! Try to eat fish seasonally from as local a source as possible. If you get the chance, ask a fisherman what they are catching at that time and what's appropriate. This helps to take the pressure off fish stocks and signals that you are only interested in sustainably caught monitored stocks that have a chance to produce offspring and regenerate the oceans.

GET TOUGH ABOUT FOOD WASTE

Research shows that planning your meals can help to cut food waste at home.[5] Proper planned meals instead of direct-from-the-fridge moments of grazing will help you use what you buy. If you do overbuy, share surplus with your friends and neighbours. There are even dedicated food share apps and platforms such as Olio.com to advertise to neighbours. Go plant-based for taste and texture. Don't just think in terms of giving up meat; look at using more plants in your cooking as an opportunity to dial up texture and taste.

Third, mix it up wherever possible. According to the global environmental NGO the WWF, 60 per cent of all calories in the human diet are provided by just three crops: rice, wheat and maize (mostly converted in our diets into high-fructose syrup). Not only does this mean we're missing many nutrients from other crops, and makes us vulnerable to pasta shortages (see page 162), but it also creates vast monocultures that undermine ecosystem health. Just like in nature, our shopping baskets should be as diverse as possible.

Fourth, swerve the processed (based on the three big agriculture crops) and go for wholefood grains, pulses and flours. The WWF has a list of fifty alternative foodstuffs to break the grip of the Big Three. You can find it at www.wwf.org.uk/sites/default/files/201902/ Knorr_Future_50_Report_FINAL_Online.pdf.

Did you know?

According to 2018 research by US campaign organization the Environmental Working Group, 70 per cent of conventionally grown fruits and vegetables in the US contain up to 230 different pesticides.[6]

So are you ready to test your knowledge and find out whether you're all set for your lunch date with the Earth? Off we go.

DO YOU KNOW YOUR ECO-ONIONS?

YOUR ESSENTIAL TEN-QUESTION QUIZ

1. **In 2021, British veg-box scheme Oddbox petitioned the UN to do what?**

A. List food waste as a new member state named 'Wasteland'
B. Impose a universal ban on the plastic packaging of vegetables
C. Reclassify food waste as an act of ecocide
D. Fine major food conglomerates that waste over 30 per cent of food in their supply chains

2. **Who dispensed the following seven words of advice, saying it summed up everything they had learned about food and health: 'Eat food, not too much, mostly plants.'**

A. Anthony Bourdain
B. Farah Quinn
C. Julia Child
D. Michael Pollan

3. It's not just what we eat, but how we eat it. What percentage of food is estimated to be eaten by people in their car in the US?

A. 10 per cent
B. 70 per cent
C. 20 per cent
D. 80 per cent

4. On New Year's Day 2022, a law making it illegal for retailers to wrap around thirty types of fruit and vegetable in plastic packaging came into force in which country?

A. France
B. Argentina
C. Kenya
D. Spain

5. The following slogan, featured on posters in the US, is a campaign for what or who? 'Keep Your Friends Close, and Your Farms Closer.'

A. The Raisin Growers Cooperative
B. Almond Farmers Union
C. Community Supported Agriculture
D. The Black Food Sovereignty Coalition

6. In previous chapters we got a sense of the incredible range of plant species hosted and nurtured by Earth. But how many different crop species are grown at scale (taken to mean at commercially significant volumes) across the world?

A. 7,000
B. 30,000
C. 170
D. 3

7. In a warming world, the milk for your breakfast cereal might have to come from a more resilient milker. Which type of cow do scientists believe that to be?

    A. Jersey
    B. Holstein
    C. Nelore
    D. Limousin

8. In 2019 a team of expert scientists from sixteen countries decided to model the Planetary Health Diet.[7] What would a diet for 10 billion people – the projected global population by 2050 – look like (and taste like!)? But what percentage of the recommended 2,500 calories per person would need to be from plant-based foods?

    A. 100 per cent – no meat or dairy can be included
    B. 90.8 per cent needs to be plant-based foods
    C. 50 per cent, with the rest split between animal protein and added sugars
    D. 12 per cent

9. A growing group of start-up companies, including Umami and Shiok, BlueNalu, Wildtype and Avant have stated they are out to disrupt environmentally high-impact industries by using labs to grow animal-free alternatives to what?

    A. Steaks
    B. Avocados
    C. Octopus
    D. Yellowfin Tuna and Red Snapper

10. Eco-art activists Daniel Fernández Pascual and Alon Schwabe form a collective, Cooking Sections. What term have they coined for a way of eating and navigating food that focuses on its impact on Earth?

    A. Veganuary
    B. Climavore
    C. Earth Calories
    D. Omnivore

· · · · · · · · · · · · · ·

ANSWERS

1. Oddbox petitioned the UN to A, list food waste as a new member state in its own right, which they suggested should be called Wasteland. There is some justification for this. If food waste was a country, its emissions would be the third-biggest contributor to global climate emissions, behind China and the US. Oddbox knows a thing or two about waste. As the company name suggests, their box schemes break away from the norm. We live in an age where, in many countries, conventional food retailers measure and grade fruit and veg, rejecting any with imperfections, leading to incredible amounts of discarded fresh produce, which becomes waste and then carbon emissions. Oddbox deals solely in the redistribution of 'wonky' produce that would not have made the grade. In my opinion, its schemes like this that deserve our support.

2. The answer is D, Michael Pollan, and the snappy advice featured in his 2008 bestselling book, *In Defense of Food*. Since then, Pollan, who is Knight Professor of Science and Environmental Journalism at the University of California at Berkeley, has written many other bestselling books, bringing

a sane approach to diet and health. He takes aim at the modern food supply chain and its impact on our health (including our gut health) as well as on the planet. He is a staunch champion of buying and consuming local food, as natural as possible, and of growing your own produce. His seven rules for eating are both memorable and instantly effective. My favourite? 'Don't eat anything that won't eventually rot.' Wise words.

3. The answer is C. Again we return to Michael Pollan for this nugget. According to this great food author, once more writing in his 2008 book *In Defense of Food*, '20 per cent of food is eaten in the car'. Over the last three decades, food consumption in industrialized economies has become synonymous with on-the-go eating. Pollan's observation sits with his advice, 'Don't buy food where you buy your gasoline.' This is the age of the Forecourt Feast, where we stock up on food – often highly packaged sugar-based snacks and fizzy or caffeinated drinks – to keep us alert while driving, and buy hydrocarbons (the compounds that make up oil) at the same time, in the same space. Take a minute to recognize how weird that is. Some research shows that putting food next to hydrocarbons is a very bad idea, as the food can absorb some of the hydrocarbon chemicals. At the very least, avoid buying olives from the petrol station – one study showed that the fat content of the oil caused the highest amount of hydrocarbon chemicals to be absorbed among all food groups.[8]

4. The law banning plastic packaging for around thirty types of fruit and veg came into force in A, France, on New Year's Day 2022. This is something many nations will need to do if they're going to turn the tide on plastic waste. An amazing 40 per cent of plastic goes into food packaging, and almost all is single-use. If you're wondering which produce is now liberated, this includes potatoes, leeks, eggplant, peppers, cucumbers, tomatoes, onions, cabbage, cauliflower, carrots,

apples, pears, kiwi fruit, plums, pineapples, mango, passion fruit, oranges, lemons and grapefruit – all must be free of plastic when sold in weights under 1.5 kilograms (3.3 pounds) in France. Cut fruit gets a free pass for now, but cherry tomatoes and green beans and even diva-ish peaches (often sold in outrageous packaging so that they can be manoeuvred through a long supply chain without attracting bruises) must be out of plastic by 2024, and berries by 2026.[9] Could this be a chance for some of the mushroom packaging we met in Chapter 4 that can potentially displace plastic to move into the fruit market?

5. The answer is C, Community Supported Agriculture (CSA). This term was coined in the mid-1980s by two European farmers, Jan Vander Tuin from Switzerland and Trauger Groh from Germany, who both moved to the US. They wanted to address the march of food produced from hulking high-impact, mono-crop farms and feedlots so large that they were practically visible from space. Conceived as an alternative distribution system for food, CSAs allow the consumer to buy a bit of the harvest in advance, giving them a stake in the farm and its bounty. This also gives smaller farmers some security and allows them a foothold in food markets that are currently dominated by faceless corporations. Many countries have a version of the CSA model. In China the model is credited with helping a shift to ecological farming, and by 2007 there were an estimated 500 farms running on CSA lines.[10] In terms of other answer options, the Black Food Sovereignty Coalition (D) addresses structural racism in food growing and increases resilience for overlooked communities.

6. The answer is C, around 170 crops are grown on a commercially significant scale. Yet 6,000 to 7,000 species have been cultivated for food and there are at least 30,000 edible plant species – this explains some of the other decoy answers. Everything we've learned in this book up until now will be

causing alarm bells in your brain. After all, we know that diversity is the key to resilience and planetary health. I'm afraid it gets even worse – out of the 170 crops grown, we depend on just 30 of them for most of our calories and nutrients. Furthermore this becomes even more selective when you account for the fact that over 40 per cent of our daily calories comes from just three staple crops: rice, wheat and maize. That's a lot of pressure to put on those harvests, not to mention a serious monoculture habit that we've inflicted on Planet Earth.[11]

7. It is indeed true that in a warming world, Jerseys (A) are lined up to be better milkers than Holsteins. As the temperature edges up, scientists are busy trying to understand how they will feed and water us (and provide us with cow juice). Researchers at Mississippi State University have extensively trawled the records of 142 Jersey and 586 Holstein cows kept by the Diary Herd Improvement Association.[12] Because of the high metabolic heat production associated with rumen fermentation and lactation, dairy cattle are particularly sensitive to heat stress. Higher temperatures lower milk output and reduce the percentages of fat, solids, lactose and protein in milk. Heat stress also reduces the fertility of dairy cows, lowering reproduction rates. But this is not going to be an easy switch. Of the more than 9 million dairy cows in the United States, 94 per cent are of Holstein descent.

Jersey (left) and Holstein (right). Which will do better in a warming world?

8. It's B, 90.8 per cent, which means a pretty huge shift in diets for most people. Avoiding meat and dairy has been shown to be the single biggest move you can make to cut your overall carbon footprint. Analysis shows that producing livestock provides just 18 per cent of calories but takes up 83 per cent of farmland globally.[13] Raising animals for meat causes twice the impact of producing plant-based foods, while, as we saw earlier, there is much talk of methane emissions from cow burps and farts (sorry!) – an unavoidable fact of life for an animal with four stomachs. What is often overlooked is the climate penalty when land is converted from forest to grazing. In a sense it is a double hit.

9. It's D: the disrupters have stated they are out to get a slice of the $1.5-trillion seafood economy with cellular seafood alternatives beginning with yellowfin tuna and red snapper (and Japanese eel). We'll keep our fingers crossed that cellular seafood takes off, because the popularity of seafood has led to chronic overfishing. Since 1961, per capita we have increased our consumption from 9 kilograms to 20.2 kilograms per person,[14] and this is expected to increase. Meanwhile, populations of marine species have halved since 1970, as we acknowledged back in Chapter 5. I hope someone is working on a version of octopus. How anyone is still eating these after watching the 2020 Oscar-winning documentary *My Octopus Teacher*, I don't know.

10. The answer is B, climavore. I like this term as it seems to me to merge changing personal food habits with food activism in an effective way. I would define being a climavore as someone who amends their diet to reflect the fact that we are in a climate and nature crisis; not just avoiding 'bad' stuff with high

emissions, but actively seeking out dietary options that will help to restore a degraded planet. For example, a climavore might swap out a problem protein such as farmed salmon for native oysters that actively regenerate their habitat and clean the waters. It's about taking your foot off the gas, using your shopping budget to send a signal to the industry to raise its standards, and causing a shift away from damaging production based on fossil fuels. For more great examples, go to becoming.climavore.org.

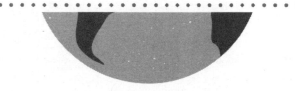

How did you do? If you scored under 3, don't panic! Remember, when it comes to amending your own behaviours you are in the driving seat. The first thing to concentrate on is where you're already doing well. If your score was middling, console yourself with the fact that you now know about a few more eco-food hacks, movements and ideas that could revolutionize your approach. If you rocked this category and proved that you are already eating for the planet, congratulations. But don't forget that this is just the start – keep pushing yourself and working out how you can continue to excel. And on that note, it's onwards to the final stop on our journey.

DESTINATION TRANSFORMATION

I hope you've enjoyed the journey through this book so far. The finish line is now in sight! Don't be too hung up on your quiz score if it's in the low numbers. You still have one round to go. Besides, it's really more important to me that you are feeling confident, inspired and included in this mission. There are so many ways of describing the complex, stressful and yet exciting thing that we're all trying to do here. I choose to express that mission in terms of striving to be the ultimate friend of the planet. For me this is a simple relatable way that has helped me when I've felt bogged down under the weight of newly published climate science reports and depressing details of failure to stick up for nature. It also seems to help other people, especially if they don't have a huge background in environmental issues. Many of us suffer from a form of imposter syndrome. We may ask ourselves, why would anybody listen to our views on the Earth, or climate or nature or consumerism, never mind act on them? Or we may be so overwhelmed by the climate and nature crisis that we struggle to know where to start. From there it's a short emotional distance to hearing a little internal voice telling you that there's nothing you can do to change things.

At various times I experience all of these fears and barriers. When I hit one of these periods of self-doubt I remind myself of a quote from the great naturalist Dr Jane Goodall: 'You cannot get through a single day without having an impact on the world around you. What you do makes a difference, and you have to decide what kind of difference you want to make.'[1] I have been lucky to work with Jane several times, and I can hear her saying these words in my mind. It sums up her pragmatism and resolve, and it makes me act.

Did you know?

The Youth Climate Strike that took place on 20 September 2019 around the world, attended by Greta Thunberg in New York, is probably the largest ever climate protest in history to date. Around 4 million children and adults came out to show their support.

Other people have developed different ways of describing the task of speaking up for the planet and making better decisions for all our sakes. Farhana Yamin is an environmental lawyer and climate change expert who has been to every major climate summit since 1991 (that means all of them!). She has advised leaders and ministers on climate negotiations for thirty years, and is particularly known for representing small islands and developing countries. In short, she's a really big deal. I recently heard Farhana speak and she is able to boil down all of her incredible knowledge into a short action point. She says that we each need to focus on becoming a JEDI. Excuse me? Farhana is using the term as an acronym. It stands for Justice, Equity, Diversity and Inclusion. So when you're thinking about and then acting on your instincts to do right by the Earth, you have to make sure that your plan considers these four fundamentals. But when you use them, you have lift-off and it's very exciting. The force is with you!

Some will try and tell you that the challenges are too great to make a difference; they might even laugh at your mission. People who do that are not usually inherently bad – they are often just forgetting to factor in the cost of failing to act and failing to change. You could, perhaps, find a polite way of telling them that the cost of catastrophic climate change and species collapse is a non-liveable planet, and because we live here, that's a high price. A few may be inherently bad, and they are best avoided.

There is no point in giving just 50 per cent in this mission. As the climate writer and podcaster Mary Annaïse Heglar puts it: 'Getting halfway out of an emergency is still an emergency. There's no middle ground when it comes to a liveable future.'[2] I interpret that as: you may as well go for it! You don't have to have all of the solutions at hand to get on with joining a group, a movement, taking a course or chatting to your friends or even neighbours about the climate and nature emergency and what we're going to do about it. But you do have to ask all the questions, so don't be fobbed off (which perhaps means also, don't be too polite). Health in our atmosphere will not be created by burning fossil fuels and the avalanche of species loss cannot be stopped if we keep developing habitats or using destructive practices in the ocean like bottom trawling. There is no point in keeping quiet about this.

CLIMATE CLASS

Many of us didn't get much nature or climate education at school. Across the world we are lacking eco-literacy, which can make it hard to articulate what a liveable planet looks like. If you feel there is a gap in your eco-education, look out for one of the thousands of online courses on climate and nature that can help get you fit for the future. MOOC stands for Massive Open Online Course and is available to as many participants as care to sign up, in many different languages. There are loads of MOOCs dedicated to the Earth, including courses in partnership with the UN Environment Programme (you get a certificate at the end of these), like this one: pedrr. org/mooc. Or you could get some incredible life-changing instruction from www.aimhi.earth. Here they not only teach you to become a great nature thinker but also plug you into one of the coolest communities of Earth defenders.

For me it is most important that we keep hopeful. I don't mean the sort of fake hopeful messages that you might find on a refrigerator magnet or a greetings card of the 'Hey, keep positive!' variety. Instead I mean the active hope that keeps you pushing for change. That sort of hope takes practice, and it needs nurturing. It needs feeding, too! One of the best ways of doing that is to keep up to date with what other humans are accomplishing. As I was writing this book and compiling the quizzes, I was constantly amazed and cheered up by how much was being achieved. Sometimes this is via noisy, bold campaigning – in the Netherlands, for example, 900 people took their government to court for reneging on climate commitments. In fact, they did it three times, winning each case.

Did you know?

Between 2010 and 2020, twenty environmental journalists were killed for covering stories on the destruction to our planet, most frequently when covering deforestation. Many more suffer violence and intimidation.

But activism can also take quieter, more specific forms. One of my favourite examples of an Earth defender on a mission is Chilean mycologist Giuliana Furci. You remember back in Chapter 4 we had a glimpse of the future of fungi (inextricably bound up with our own future it seems). In 2021, thanks to her tireless campaigning and work uncovering the fundamental ecological importance of fungi, the Species Survival Commission (SSC) of the International Union for Conservation of Nature (i.e. the most important human body on Earth for nature) announced the addition of a third kingdom classification. They would, it announced, add a third 'F' to the existing kingdom classification duo of flora and fauna? This gives us flora, fauna and funga. That's huge!

SOUNDS OF THE UNDERGROUND

If you want to get more involved with that third 'F' for funga, SPUN is the Society for the Protection of Underground Networks, a giant collaboration between researchers and local communities that will map fungal networks and protect them. SPUN frequently calls on citizen scientists (those of us armed with our phones and instructions) to help feed in their observations to help with the mapping. That will give a new sense of purpose to your forest hikes!

On that note of optimism, are you ready for your final round? This last challenge is designed to uplift and inspire you. We'll meet some bona fide ultimate friends and we'll find out that change isn't just possible, but is happening at scale and speed.

ONE LAST CHANCE TO STEP UP OR STEP OFF

YOUR ESSENTIAL TEN-QUESTION QUIZ

1. How did environmental visionary Wangari Maathai celebrate on hearing the news that she had been awarded the Nobel Peace Prize?

A. By going out to dinner with Oprah Winfrey
B. By planting a tree
C. By taking part in a beach clean
D. By leading an expedition to a polar research station

2. In her famous song 'Big Yellow Taxi', how much does Joni Mitchell say the entrance fee is to see the trees in the tree museum?

A. A dollar and a half
B. Zero (because there are no trees to see any more)
C. 25 bucks
D. 50 cents

3. Here's another question from popular culture. The dogged environmental activism of former legal clerk Erin Brockovich was portrayed by Julia Roberts in a famous film of the same name (note the real Erin now goes by the name Erin Brockovich-Ellis). She led a direct-action suit against a company accused of polluting the water supply of an entire town in southern California. But what was the name of the company she squared up to?

A. Pasedena Pipelines and Utilities
B. Hinkley Compressors
C. The Pacific Gas and Electric Company
D. Burns Slant Drilling

4. Let's not forget that some of our greatest inspirational allies of Earth are fictional. Take this one for example: 'Explorer, warrior and peacemaker on her journey to end conflict between human factions in a world where humanity is constantly beset by a hostile environment.'[3] Which revered fictional character does this refer to?

A. Lara Croft, *Tomb Raider*
B. Nausicaä of the Valley of the Wind
C. Kate Dibiasky in the hit Netflix film *Don't Look Up*
D. Valkyrie, Marvel's Asgardian Warrior

5. On 1 May (also known as May Day in the UK) in 2019, what might UK citizens boiling water in their kettles to make their fabled and much-loved cups of tea have missed?

A. That they were using electricity generated without burning coal for the first time since the Victorian era
B. That all electricity that day was generated by solar, thanks to a historic sunny day
C. That the kettle took twice as long to boil as the UK grid halved output to commemorate the 42nd global Earth Day
D. That energy was created by nuclear fusion for the first time

6. In 2015 Pope Francis released an encyclical (the highest form of papal document that can be released by a pope alone) with the subtitle 'on care for our common home'. What is the name of this Earth-focused document?

    A. *Terra Carta*
    B. *Laudato Si'*
    C. *One Planet 'Marshall Plan'*
    D. *Stella Maris*

7. According to the founder of a huge grassroots project, Afroz Shah, 'The whole edifice of this movement is based on love.' But what Earth-defending act is Afroz famed for?

    A. The world's biggest beach clean
    B. The world's biggest fossil-fuel protest
    C. The world's biggest tree-planting project
    D. The world's biggest clothes-mending workshop

8. The endurance swimmer and ocean campaigner Lewis Pugh has become synonymous with death-defying swimming challenges that draw global attention to the ecological crisis. What has his brand of activism become known as?

    A. Human icebreaking
    B. Wetsuit wrangling
    C. Speedo diplomacy
    D. Backstroke bargaining

9. We're starting to think about economics differently. Professor of economics and bestselling author Kate Raworth has created and defined a system of Earth-friendly economics that meets the needs of all while protecting the Earth. But what is its delicious name?

A. Fair pie theory

B. Cake slice

C. Choc-chip cookie theory

D. Doughnut economics

10. It's back to Earth for the last question. So, here goes: what is the name of the system that allows natural processes to shape land and ocean, repair damaged ecosystems and restore degraded landscapes?

A. Earth engineering

B. Rewilding

C. Biodynamics

D. Homesteading

· · · · · · · · · · · · · · · ·

ANSWERS

1. The answer is B. Perhaps not surprisingly, the woman who launched a tree-planting revolution planted a tree by way of celebration! History records that it was a Nandi flame tree, native to her home region of Nyeri, in the central highlands of Kenya.[4] Wangari Maathai began to plant trees in the mid-1970s with women in rural communities in Kenya. These were women in poor communities, neglected by other environmental movements and projects, but on the frontline of environmental destruction and impact. The Green Belt Movement she created is the very definition of a grassroots movement, mobilizing thousands of women

whom she referred to as 'foresters without diplomas'. They could earn income by selling seedlings for reforestation and by planting trees to replenish the soil, protect water sources and grow fruit. By the time of her death in 2011, Wangari's movement had planted 30 million trees in Kenya. Globally, the Green Belt Movement has led to 11 billion trees being planted. Incidentally, Wangari Maathai was indeed friends with Oprah Winfrey, who presented her with the Nobel Peace Prize and remembered her friend after her death as 'one of the great ones; a giant sequoia in the forest of humanity.'[5]

2. The answer is A, a dollar and a half. The inspiration for one of Joni's best-loved songs apparently came from a 1969 trip she took to Hawaii. She remembers viewing a picturesque scene from her balcony and then leaning out a little further and seeing an ugly concrete car park. On the same trip she noticed the island's botanical gardens were charging a dollar and a half entrance price. (FYI, Foster's Botanical Gardens now charges $5 for an adult admission.[6]) If you want to be a true eco-king/queen you will need to know the words to this environmental anthem. Part of its brilliance lies in the way Joni Mitchell blends the personal and the political with a sense of nostalgia for what is already lost. In a later jazz-style version, she changed the admission price to an 'arm and a leg'. If you chose C, you may have heard artist Amy Grant's version, as this is the entrance fee she sings about.

3. The company Erin Brockovich squared up to so courageously was C, the Pacific Gas and Electric Company. They eventually settled in 1996 to the tune of $333 million. Interviewed twenty years on from the release of the movie, Erin Brockovich-Ellis said, 'We're still talking about this, and that movie came out twenty-one years ago. And it was almost before its time because it was about the environmental pollution that we are in the throes of today.'[7] For me

Brockovich was especially important because she redefined what it means to be an environmental activist. She showed that you need guts and determination to see it through, and that you don't have to be born into certain circumstances. Today she still works as an advocate for the planet and has inspired thousands of others to stand up to environmental destruction too, from all backgrounds and circumstances.

4. The answer is B, Nausicaä of the Valley of the Wind, the heroine of an eponymous manga and film by Hayao Miyazaki. It was made in the 1980s but has arguably become increasingly relevant as we look for visual ways of describing a collective future in which we repair our broken Earth–human relationship and the enemies that stand in our way. (If you have access to Netflix, you can watch the film *Nausicaä*, or look out for fan screenings at cinemas across the world. It is well worth watching.) *Nausicaä of the Valley of the Wind*, alongside other work by Miyazaki, is often defined as 'solar punk', a visual language to describe this great transformation to a green fair planet where the Earth is central to our decisions. For more inspiration (especially if you're somebody who responds to the visual arts) head to the platform threetransitions.earth created by Japanese anthropologists Koji Sasaki and Yosuke Ushigome of design innovation studio Takram with Hitachi. They use similar beautiful visual language to show how we might transition from the old world to the new healthy biosphere.

5. The correct answer is A: 1 May 2019 marked a historic week, the first coal-free week since 1882 when a plant opened in Holborn, London. Since 1882, in common with many industrialized nations, the UK has had a big weakness for coal. But coal-free days are becoming increasingly common as renewables begin to take some of the strain off the national grid, as part of the programme to phase out coal-fired electricity production by 2025.

6. The answer is B, *Laudato Si'*. Written to be read and acted upon by the whole of the Catholic Church, the encyclical was notable for its strength, calling out climate change as a manmade phenomenon and commenting on consumerism and irresponsible development. In it, the Pope calls on all people of the world to take 'swift and unified global action'. If you answered A, *Terra Carta*, that was the name given to the ten-point pandemic recovery charter issued by the heir to the UK throne, the Prince of Wales. A long-time environmentalist, HRH Prince Charles challenged businesses in the *Terra Carta* to 'put nature, people and the planet at the heart of global value creation'.[8]

7. The answer is A, the world's biggest beach clean. When a young Mumbai lawyer, Afroz Shah, moved into an apartment overlooking the city's Versova beach, he was horrified by the plastic pollution covering the shoreline. In 2015 he decided to gather local residents and begin a beach clean at Versova, a 2.4-kilometre strip of coastline facing the Arabian Sea. One thing led to another and a year later Afroz had mobilized thousands of people at Versova, making it the biggest beach clean in the world. Together they removed millions of kilos of plastic and packaging pollution and their influence has spread across the world as millions have begun attending regular clean-ups inland and on the coast. At Versova in 2018, their work paid off. Eighty endangered Olive Ridley sea turtle hatchlings were spotted heading towards the sea. For the first time in a decade the turtles were back in force. As we read in the question, Afroz attributes all of this to love: 'The moment we start loving other human beings we start loving nature at the same time.'

8. It's C, Speedo diplomacy. Lewis Pugh, a former marine lawyer turned endurance swimmer and ocean activist, is the first person to complete a long-distance swim in every ocean of the world, recently completing a multi-day swim in the polar

regions. During these challenges, which he does in swimming trunks (rather than a wet or dry suit) he is making the meta-point that it should not be possible for him to swim through these areas. They should be covered in thick ice. It is only due to global heating that he can make his attempts. The alliance of courage and physical prowess in Pugh's swims seem to get through where other attempts to start conversations on climate have failed. His challenges seem to wake up both global leaders and global audiences to issues such as ice loss at the Arctic and Antarctic. So yes, a human icebreaker (if you picked A), but it's all about the Speedos!

There are two very important things to bear in mind when it comes to ocean activism. Beach cleans are *not* just litter picks; and the ocean needs you – even if you can't see it out of your window. The data gleaned from the mobilization of global citizens, particularly on the volume of single-use plastics that they have collected, has helped to force legislators to take action on plastic. I feel strongly that whoever you are and wherever you live you should get involved in ocean advocacy; yes, even if you live in a landlocked country! You don't even have to wear Speedos or hold any diving records. As we've seen throughout this book, our fate is inextricably entwined with the fate of the ocean.

One of the best things that you can do is to support the ocean by becoming a member of an NGO (non-governmental organization). There are many global ocean charities to choose from, including surfrider.org, cleanseas.org and Lewispughfoundation.org, or you might choose to support a local or national charity. I am a long-term supporter of Surfers Against Sewage – sas.org.uk – a UK-based ocean advocacy charity that works not just on plastic but also on sewage (as the name suggests; water quality is a really big problem for us in the UK), as well as on overfishing and all areas that impact on the seas. I promise you I'm not a good surfer and you don't have to even get into the water to support them. Although they are a national charity, they have an international network of ocean supporters, and I help to lend my voice alongside thousands of other supporters to the global push to protect 30 per cent of our ocean by 2030 (30x30). Anything you can do to help grow this drumbeat to a deafening clamour is worth it!

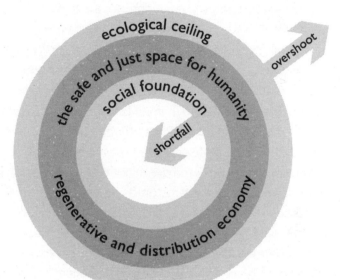

For fairer finance think of a doughnut with different fillings.

9. The answer is D, doughnut economics. While studying conventional economics (which I think it is fair to say she did not enjoy!), the British academic and innovator Kate Raworth had a major lightbulb moment. She realized that economics has lost its purpose. As she saw it, the pursuit of 'growth' pushes society into inequality and that in turn pushes us all into ecological collapse. You can see her point. As we've seen at different points in our journey through this book, we have already overshot on many of the Earth's critical biosystems. Yet still we're failing to meet the basic needs of millions of inhabitants. So Kate set about designing a different system that meets the needs of all and protects the Earth and its inhabitants. She calls this the 'sweet spot', and that's part of

the reason for the name: doughnut economics. Following the principles of doughnut economics allows us to develop a mindset that works with planetary boundaries (see page 191 for this outer bit of the circle), and therefore gives us a fighting chance of not just surviving but also thriving – that's the hole in the middle. These principles are gaining traction across the world.

10. The answer is B, rewilding. If the Earth did endorsements, I think rewilding would get a huge thumbs-up. Rewilding leads to the restoration of natural landscapes that have been damaged by human activities. It's officially one of the most efficacious and cost-effective responses to the climate and nature crisis. Scientists estimate that if one third of Earth's most degraded areas was restored and we boosted the protection of areas that are still in reasonably good condition then we could store half of all greenhouse gas emissions and prevent the loss of around 70 per cent of predicted extinctions.[9] There are many ways of supporting the rewilding movement, and one of the most effective is by tapping into rewilding in your country or possibly in your local area.

THE STARS OF EARTH ACTIVISM

Re:wild is the organization co-founded by the actor and climate campaigner Leonardo DiCaprio. It's quickly become one of the biggest players in rewilding, collaborating with scientists, NGOs and indigenous communities to focus on restoring hot spots of biodiversity, including in the Galapagos Islands. Re:wild is a 'force multiplier', which means it amplifies and quickens all the existing efforts. It gives me a lot of hope, because I get very excited when we all come together to amplify and boost our work. Re:wild also acknowledges that a huge proportion of Earth-defending is taken on by indigenous communities. In fact, all over the world indigenous people are on the frontline of defending the nature hot spots and carbon sinks that we all depend on, including the rainforests in the Amazon and the Congo. But indigenous Earth defenders are more likely to be attacked and killed. Globalwitness.org helps to defend the defenders. As DiCaprio puts it, 'The environmental heroes that the planet needs are already here. Now we all must rise to the challenge and join them.'[10]

Can you believe it – you've completed the final chapter and the final quiz. The best transformations rarely happen overnight and usually involve baby steps to giant leaps forward. So if you still have a few steps to take, don't worry! If your score was low in this final round, it's probably because you were meeting some ideas and concepts for the first time. You just need to build confidence and put some of the theory you've hopefully picked up into action. If you romped across the finish line to this last chapter with a big score (7+) that's a sign that your next step should be to grow your individual commitments and interest into a bigger communal effort. Ask yourself which existing organization you can join to amplify your passion and commitment to an Earth-friendly life. And if that organization doesn't yet exist, well, can you start it? I'm really excited to see where this can take you. But for now, it's time to tot up our overall scores and draw some big conclusions.

THE FINAL COUNTDOWN
HOW DID YOU SCORE?

Now we get to the very end. It's the moment of truth when we tot up your overall score and find out where you have landed. Whatever that score is, let me be clear that the very act of picking up this book and having a go at ten different quiz sections means you are a true friend of the planet. Remember, if your score is on the low side on your first attempt, it just means that you have the opportunity of learning more. The more you learn and absorb, the better friend you can be to the Earth!

1–20:

A valiant attempt. Perhaps you were new to the idea of sustainability – which means being a custodian rather than owner of the biosphere – and to the idea of handing a healthy planet on to future generations. You've come across some big themes and ideas for the first time, so all credit to you for having a go. Once you've picked up some of the main ideas such as prioritizing the Earth in decision-making and had a chance to try them out in real life, there will be no stopping you. Starting out with a low score gives you the opportunity to accelerate quickly, developing a whole new Earth-first mindset.

21–40:

Well done, this is solid ground. You've got some of the basics and despite the noise and distractions of everyday life, you are already taking care to pick up some important information about earth science, sustainable living and future strategies for living more carefully in harmony with the planet. Now you need to nurture those instincts. Perhaps you are tuned in to specific parts of the climate and nature emergency; for example, you might have felt very strongly about the amount of plastic waste littering the globe. Your task now is to join the dots. For instance, plastic waste is made from oil and that's a fossil fuel and driving climate change. Follow up with some of the resources I list in this book, and you will soon be able to do just that.

41–60:

Mid-place in our scoring means you have developed well-rounded knowledge of the way the planet works and you are on the cusp of being a very useful ally. Now all you need to do is to turn up the dial on your interest in the biosphere. You can do this by doubling your reading/scrolling and conversations about earth systems and sustainability. But even better would be for you to join an NGO or charity focused on a major Earth-friendly objective, whether that be climate, ocean health or ending deforestation (I hope you've picked up some ideas throughout this book). It's time to get focused and turn your skills into laser-focused activism!

61–80:

I am proud that you are so connected to the workings of the biosphere and so knowledgeable about the history, culture and, dare I say, the future of environmentalism. With one last push you can catapult yourself into the top league. Identify your weakest category and see if you can build your knowledge there. You're very nearly an invincible planetary ally!

81–100:

Congratulations. You are gloriously and unequivocally a true friend to this planet. You have put the time in to learn as much as you can about the workings of ecosystems and how you and your fellow humans can not only limit your impact, but also help the planet return to health. You are in step with the Earth and a real influencer when it comes to green behaviours and knowledge. Now your task is to increase your sphere of influence and persuade as many of your friends, family and followers as you can (if you use social media) to strive to be an ultimate friend of the Earth like you. But don't let your status go to your head. Staying humble and putting the Earth first has got you to this position, so remember to keep it that way.

CONCLUSION

YOU'VE REACHED THE END OF THE EARTH (SAVING)

However you did, I want to thank you for rising to this challenge. I'm hoping this book is just the start and leads you on to as much Earth activism and championing as you can cram into your days (and maybe nights!). Feel free to recycle and reuse this book as many times as you can.

It can be daunting to think of yourself as an agent of change (which sounds quite grandiose in and of itself). But as any seasoned activist will tell you, everyone has a sphere of influence through which they can inspire people by their actions. It is also more important to do stuff than to be seen to do stuff, so don't worry at all if you hate social media and are photophobic – the most industrious and effective Earth-defenders are often the quietest ones, who don't spend all day taking selfies. Neither do you need to be out on the savannahs or jumping down gorges to make an impact. I know this is quite the name-drop, right at the end of a book, but during the first lockdown in England at the height of the global pandemic, I was lucky enough to interview Sir David Attenborough by phone for our podcast, So Hot Right Now, on how we can tell better stories about nature and climate. My favourite bit was when Sir David began to talk about the birds that he could see in his London garden. The great natural history storyteller had had his own wings clipped by the global pandemic (he would usually be in some far-flung location), but seemed to derive as much interest and pleasure in London birds near to Richmond Park as in observing creatures in the Galapagos Islands.

That was a lesson for me too. We are all lucky today to have so much information and science to pore over.

We have a lot on our shoulders, but we have a lot on our side. Congratulations, you have just become the ultimate friend of the Earth.

ENDNOTES

Chapter 1

1   'Biosphere', Resource Library, National Geographic, www.nationalgeographic.org/encyclopedia/biosphere/.

2   Anesio, A, and Laybourn-Parry, J., University of Bristol, 'Glaciers and Ice Sheets as a Biome: Trends in Ecological Evolution' (2012), pubmed.ncbi.nlm.nih.gov/22000675/.

3   'Climate Change on the Third Pole', University of Aberdeen Working Paper (2021), ed6ab6c8-ea4c-4ae8-92aa-61d3cd1bebb.usrfiles.com/ugd/ed6ab6_31104d85928146ae9a6d73cc99a26834.pdf.

4   The Blue Marble, NASA, www.nasa.gov/image-feature/the-blue-marble-the-view-from-apollo-17.

5   Soloman, Christopher, 'Life in the Balance', *National Geographic* (June 2017).

6   Teff-Skekker, Y. and Orenstein, D., 'The Desert Experience', British Ecological Society Research Paper (2019), besjournals.onlinelibrary.wiley.com/doi/full/10.1002/pan3.28.

7   Vidal, J., 'Scientists Watch Giant "Doomsday" Glacier in Antarctica with Concern', *Guardian* (December 2021), www.theguardian.com/world/2021/dec/18/scientists-watch-giant-doomsday-glacier-in-antarctica-with-concern.

8   World Atlas, 'Countries Sharing the Amazon Rainforest', www.worldatlas.com/articles/countries-sharing-the-amazon-rainforest.html.

Chapter 2

1   'How Oman's Rocks Could Help Save the Planet', *New York Times Magazine*, www.nytimes.com/interactive/2018/04/26/climate/oman-rocks.html.

2   'Planetary Boundaries', Stockholm Resilience Centre, www.stockholmresilience.org/research/planetary-boundaries.html.

3   www.theguardian.com/environment/2021/jan/07/global-heating-stabilize-net-zero-emissions.

4   Cutzen, Paul, 'Geology of Manking', *Nature* (2002), www.nature.com/articles/415023a..

5   'The Carbon Inequality Era', Research Paper, Stockholm Environment Institute and Oxfam (2020), www.sei.org/wp-content/uploads/2020/09/research-report-carbon-inequality-era.pdf.

6   Godfrey-Smith, P., 'The Ant and the Steam Engine', *London Review of Books*, Vol. 37, No. 4 (Feb. 2015), www.lrb.co.uk/the-paper/v37/n04/peter-godfrey-smith/the-ant-and-the-steam-engine.

7   Bielo, D., '400 PPM: Carbon Dioxide in the Atmosphere Reaches Prehistoric Levels', *Scientific American* (2013), blogs.scientificamerican.com/observations/400-ppm-carbon-dioxide-in-the-atmosphere-reaches-prehistoric-levels/.

8   Which is a Bigger Methane Source: Cow Belching or Cow Flatulence?', *Global Climate Change*, NASA, climate.nasa.gov/faq/33/which-is-a-bigger-methane-source-cow-belching-or-cow-flatulence/.

9   Gaukel Andrews, C., 'More Animals and More Biodiversity Mean More CO_2 Storing', *Good Nature Travel* (2018), www.nathab.com/blog/biodiversity-matters-the-more-large-animals-the-more-co2-storing/.

10   'Carbon Dioxide Emissions from Energy Consumption in the United States from 1975 to 2020', *Statista*, www.statista.com/statistics/183943/us-carbon-dioxide-emissions-from-1999/.

11   Greta Thunberg, 'COP26 even watered down the blah, blah, blah', BBC News (November 2021), www.bbc.co.uk/news/av/uk-scotland-59298344.

12   'Subtitles to Save the World – 2021', BAFTA Albert, wearealbert.org/editorial/wp-content/uploads/sites/6/2021/09/albert-subtitle-report-2021.pdf.

Chapter 3

1   Bridge, H., 'Earthworms' Place on Earth Mapped', *BBC News* (Oct. 2019), www.bbc.co.uk/news/science-environment-50157313.

2   Cavan, E. L., Belcher, A., Atkinson, A., et al., 'The Importance of Antarctic Krill in Biogeochemical Cycles, *Nature Communications* (Oct. 2019), www.nature.com/articles/s41467-019-12668-7.

3   'Well, this is the million dollar question...', WWF, wwf.panda.org/discover/our_focus/biodiversity/biodiversity/

4   Mitchell, A., '"Earthworm Dilemma" Has Climate Scientists Racing to Keep Up'., *New York Times* (May 2019), www.nytimes.com/2019/05/20/science/earthworms-soil-climate.html.

5   '22 Animals That Went Extinct in the US in 2021', *Global Citizen*, www.globalcitizen.org/en/content/animal-extinct-biodiversity-2021/.

6   'Bird Classification', RSPB, www.rspb.org.uk/birds-and-wildlife/natures-home-magazine/birds-and-wildlife-articles/how-do-birds-survive/bird-classification/.

7   Waters, H., 'New Study Doubles the World's Number of Bird Species by Redefining "Species"', *Audubon* (2016), www.audubon.org/news/new-study-doubles-worlds-number-bird-species-redefining-species.

8   'A Peek into the US Tiger Trade', *National Geographic* (Dec. 2019).

9   'Ultimate Hunter', *National Geographic* (Aug. 2017).

10   Glenza, J., 'Florida Will Begin Emergency Feeding and Rescue of Starving Manatees', *Guardian* (2016), www.theguardian.com/environment/2021/dec/10/florida-manatees-rescue-emergency-feeding..

11   Chesnes, M., 'Manatee feeding was to begin Wednesday. The only thing missing? Manatees', *Treasure Coast News*, eu.tcpalm.com/story/news/local/indian-river-lagoon/2021/12/15/manatee-feeding-program-not-exactly-off-great-start-says-fwri/8906881002/.

12   Maron, Dina Fine, 'How the World's Largest Rhino Population Dropped by 70 Percent – in a Decade, *National Geographic* (2021), www.nationalgeographic.com/animals/article/rhino-numbers-drop-kruger-national-park.

13   'Rhino Facts', WWF, www.worldwildlife.org/species/rhino.

14   'Croaking Science', *Frog Life* (2020), www.froglife.org/2020/11/29/croaking-science-how-many-amphibian-species-are-there-how-do-we-know-and-how-many-are-threatened-with-extinction/.

15   Carrington, D., 'A Car "Splatometer" Study Finds Huge Insect Die-Off', *Wired* (2020), www.wired.com/story/a-car-splatometer-study-finds-huge-insect-die-off/.

16   Ibid..

17   'How many species on Earth? Why that's a simple question but hard to answer', *The Conversation* (2020), theconversation.com/how-many-species-on-earth-why-thats-a-simple-question-but-hard-to-answer-114909.

Chapter 4

1   Harris, N., and Gibbs, D., 'Forests Absorb Twice as Much Carbon as They Emit Each Year', World Resources Institute (2021), www.wri.org/insights/forests-absorb-twice-much-carbon-they-emit-each-year.

2   www.tenmilliontrees.org/trees.

3   Leahy, S., 'Half of All Land Must Be Kept in a Natural State to Protect Earth', *National Geographic* (2019), www.nationalgeographic.com/environment/article/science-study-outlines-30-percent-conservation-2030.

4   Weisse, M., 'We Lost a Football Pitch of Primary Rainforest Every 6 Seconds in 2019', World Resources Institute (2020), www.wri.org/insights/we-lost-football-pitch-primary-rainforest-every-6-seconds-2019.

5   Watson, T., 'Wood You Believe It? Earth has 3 Trillion Trees!', *USA Today*, eu.usatoday.com/story/news/2015/09/02/earth-three-trillion-trees/71578324/.

6   'Russia's Forests Store More Carbon Than Previously Thought', European Space Agency, www.esa.int/Applications/Observing_the_Earth/Space_for_our_climate/Russia_s_forests_store_more_carbon_than_previously_thought.

7   Ibid.

8   Ibid.

9   'Species-rich Forests Store Twice as Much Carbon as Monocultures', University of Zurich, www.sciencedaily.com/releases/2018/10/181004143905.htm.

10   Zabarenko, D., 'Hurricane Katrina boosted greenhouse gases-report', Reuters (2007), www.reuters.com/article/katrina-carbon-idUSN1243324320071115.

11   'Facts and Figures', Air Transport Action Group, www.atag.org/facts-figures.html.

12   Gammon, K., 'How the billionaire space race could be one giant leap for pollution', *Guardian* (2021), www.theguardian.com/science/2021/jul/19/billionaires-space-tourism-environment-emissions.

13   'Rainforest Facts', Rainforest Maker, www.rainforestmaker.org/facts.html.

14   Loomis, I., 'Trees in the Amazon Make Their Own Rain', *Science* (2017), www.science.org/content/article/trees-amazon-make-their-own-rain.

15   Schiffman, R., '"Mother Trees" Are Intelligent: They Learn and Remember, *Scientific American* (2021), www.scientificamerican.com/article/mother-trees-are-intelligent-they-learn-and-remember/.

16 'Source For All: Summer 2022', Mylo Frayme release, PDF, Stella McCartney. Plus author interviews, SM and Bolt Threads.

Chapter 5

1 Watson, A. J., Schuster, U., Shutler, J. D., et al. 'Revised Estimates of Ocean-Atmosphere CO_2 Flux Are Consistent With Ocean Carbon Inventory', *Nature Communications*, vol 11. no. 4422 (2020) www.nature.com/articles/s41467-020-18203-3#Sec2.

2 'Why Should We Care About the Ocean?', NOAA, oceanservice.noaa.gov/facts/why-care-about-ocean.html#:~:text=The%20air%20we%20breathe%3A%20The,our%20climate%20and%20weather%20patterns.

3 Original quote, Interview, *OMNI Magazine* (July 1992), p. 66.

4 'The Sixth Status of Corals of the World: 2020 Report', Global Coral Reef Monitoring Programme, gcrmn.net/2020-report/.

5 'The Ocean, A Carbon Sink', Ocean and Climate Platform, ocean-climate.org/en/awareness/the-ocean-a-carbon-sink/.

6 *Ocean*, National Geographic Resource Library, www.nationalgeographic.org/encyclopedia/ocean/.

7 Glancy, J., 'No Man's Water', *Oceangraphic Magazine* (2021), www.oceanographicmagazine.com/features/shark-survival-palau/.

8 'Overfishing Puts More Than One Third of All Sharks, Rays, and Chimaeras at Risk of Extinction', WWF (2021), www.worldwildlife.org/stories/overfishing-puts-more-than-one-third-of-all-sharks-rays-and-chimaeras-at-risk-of-extinction.

9 Earle, Sylvia, 'National Geographic Ocean: A Global Odyssey', *National Geographic* (2021).

10 Sala, E., Mayorga, J., Bradley, D., et al. 'Protecting the Global Ocean for Biodiversity, Food and Climate. *Nature* 592 (2021), www.nature.com/articles/s41586-021-03371-z.

11 Ritchie, Hannah, and Roser, Max (2021), 'Biodiversity', ourworldindata.org/biodiversity.

12 McKie, R., 'Is Deep Sea Mining a Cure for the Climate Crisis or a Curse?', O*bserver* (2021), www.theguardian.com/world/2021/aug/29/is-deep-sea-mining-a-cure-for-the-climate-crisis-or-a-curse.

13 Milko, V., 'Rare, Pristine Coral Reef Found off Tahiti Coast', Associated

Press 2022, lasvegassun.com/news/2022/jan/19/rare-pristine-coral-reef-found-off-tahiti-coast/.

14   'All About the Ocean', National Geographic Resource Library, www.nationalgeographic.org/article/all-about-the-ocean/.

15   'Plastic Bag Found at the Bottom of World's Deepest Ocean Trench', *National Geographic*, www.nationalgeographic.org/article/plastic-bag-found-bottom-worlds-deepest-ocean-trench/.

Chapter 6

1   Report: '"Throwaway Global Economy" is Fuelling Climate Change', Circular (2022), www.circularonline.co.uk/news/report-throwaway-global-economy-is-fuelling-climate-change/.

2   Pochin, Courtney, 'Model "gutted" as £18 Missguided jumpsuit "ruins" her new £60k Porsche', *Mirror* (13 Nov. 2020), www.mirror.co.uk/news/uk-news/model-gutted-18-missguided-jumpsuit-23003587

3   'Researchers Use Brain Scans to Predict When People Will Buy Products', Carnegie Mellon University:, www.cmu.edu/news/archive/2007/January/jan3_brainscans.shtml.

4   Weise, K., 'Jeff Bezos Commits $10 Billion to Address Climate Change', *New York Times* (published 2020, updated 2021), www.nytimes.com/2020/02/17/technology/jeff-bezos-climate-change-earth-fund.html.

5   Ibid.

6   'Waste and Pollution, Clean Clothes Campaign', cleanclothes.org/fashions-problems/waste-and-pollution.

7   'Our Addiction to Plastic', *National Geographic* (December, 2019 print edition).

8   'Moh's Scale of Hardness', Collector's Corner, www.minsocam.org/msa/collectors_corner/article/mohs.htm.

9   Weise, K, 'Jeff Bezos Commits $10 Billion to Address Climate Change', *New York Times* (published 2020, updated 2021), www.nytimes.com/2020/02/17/technology/jeff-bezos-climate-change-earth-fund.html.

10   Long, K. A., 'Amazon settles with two Seattle workers who say they were wrongfully fired for their advocacy', *Seattle Times* (2021), www.seattletimes.com/business/amazon/amazon-settles-with-two-seattle-workers-who-say-they-were-wrongfully-fired-for-their-advocacy/

Chapter 7

1   'Age of the Earth', National Geographic Resource Library, www.nationalgeographic.org/topics/resource-library-age-earth/?q=&page=1&per_page=25.

2   'Safe planetary boundary for pollutants, including plastics, exceeded, say researchers', Stockholm Environment Institute (2022), www.sei.org/about-sei/press-room/safe-planetary-boundary-pollutants-plastics-exceeded/.

3   Ibid.

4   Ibid.

5   'Use and Reuse', info site, Levi Strauss, www.levistrauss.com/how-we-do-business/use-and-reuse/.

6   'River Thames: Mounds of Wet Wipes Reshaping Waterway', *BBC News* (2021), www.bbc.co.uk/news/uk-england-london-58742161.

7   Wetzel, C., 'This New Installation Pulled 20,000 Pounds of Plastic from the Great Pacific Garbage Patch', *Smithsonian* (2021), www.smithsonianmag.com/smart-news/this-new-installation-just-pulled-20000-pounds-of-plastic-from-the-great-pacific-garbage-patch-180978895/.

8   'Space Debris and Human Spacecraft', NASA (2021), www.nasa.gov/mission_pages/station/news/orbital_debris.html.

9   'What Happens to NYC's 3.2 Million Tons of Trash', *Business Insider* (2021), www.businessinsider.com/what-happens-to-new-york-city-trash-2021-3?r=US&IR=T.

10   Lippard, L., 'New York Comes Clean: The Controversial Story of the Fresh Kills Dumpsite', *Guardian* (2016), www.theguardian.com/cities/2016/oct/28/new-york-comes-clean-fresh-kills-staten-island-notorious-dumpsite.

11   Lebreton, L., Slat, B., Sainte-Rose, J., et al., 'Evidence that the Great Pacific Garbage Patch is Rapidly Accumulating Plastic', Scientific Reports – *Nature*, vol 8., No. 4666 (2018).

12   Laville, S., 'A Million Bottles a Minute: World's Plastic Binge "as Dangerous as Climate Change"', *Guardian* (2017), www.theguardian.com/environment/2017/jun/28/a-million-a-minute-worlds-plastic-bottle-binge-as-dangerous-as-climate-change.

13   Ibid.

14   Greenpeace Canada press release (2018), www.greenpeace.org/canada/en/press-release/277/press-release-greenpeace-slams-coca-cola-plastic-announcement-as-dodging-the-main-issue/.

15   McVeigh, K, 'Nurdles: The Worst Toxic Waste You've Probably Never Heard Of', *Guardian* (2021), www.theguardian.com/environment/2021/nov/29/nurdles-plastic-pellets-environmental-ocean-spills-toxic-waste-not-classified-hazardous.

16   'X-Press Pearl Maritime Disaster: Sri Lanka', UNEP report (2021), postconflict. unep.ch/Sri%20Lanka/X-Press_Sri%20Lanka_UNEP_27.07.2021_s.pdf.

17   Hartline, N. L., 'Microfiber Masses Recovered from Conventional Machine Washing of New or Aged Garments', ACS Publications, pubs.acs.org/doi/abs/10.1021/acs.est.6b03045.

Chapter 8

1   CNBC, 'Smartphone Users Are Waiting Longer Before Upgrading – here's why' (2019), www.cnbc.com/2019/05/17/smartphone-users-are-waiting-longer-before-upgrading-heres-why.html.

2   Fairtrade foundation fact sheet, www.fairtrade.org.uk/media-centre/blog/top-12-facts-about-fairtrade-bananas/.

3   Bodyflik, Museum of Design in Plastic, www.modip.ac.uk/artefact/aibdc-005910.

4   '1.5 Degree Lifestyles', Hot or Cool Institute, report, hotorcool.org/wp-content/uploads/2021/01/15_Degree_Lifestyles_MainReport.pdf.

5   'How the Rich Are Driving Climate Change', BBC Climate, www.bbc.com/future/article/20211025-climate-how-to-make-the-rich-pay-for-their-carbon-emissions.

6   Thunberg, G., '"Our House is on Fire": Greta Thunberg, 16, Urges Leaders to Act on Climate', *Guardian* (2019), www.theguardian.com/environment/2019/jan/25/our-house-is-on-fire-greta-thunberg16-urges-leaders-to-act-on-climate.

7   Wortham, J., 'A Netflix Model for Haute Couture', *The New York Times* (2009), www.nytimes.com/2009/11/09/technology/09runway.html.

8   iFixit website, www.ifixit.com/Right-to-Repair/Intro.

9   'Single-use Beverage Cups and Their Alternatives', UNEP report, www.lifecycleinitiative.org/wp-content/uploads/2021/02/UNEP_-LCA-Beverage-Cups-Report_Web.pdf.

10   'Christmas Trees: Real or Fake?', *BBC News* (2016), www.bbc.co.uk/news/uk-england-38129835.

11   Ibid.

Chapter 9

1 'Worldwide Food Waste', UNEP, www.unep.org/thinkeatsave/get-informed/worldwide-food-waste.

2 'Food Waste Bad Taste', Sustainable Restaurant Association, thesra.org/wp-content/uploads/2019/09/Food-Waste-Bad-Taste-Intro-Slides-for-Download.pdf.

3 Xu, X., Sharma, P., Shu, S., et al., 'Global Greenhouse Gas Emissions from Animal-Based Foods Are Twice Those of Plant-Based Foods, *Nature Foods*, 2 (2021), 724–32, www.nature.com/articles/s43016-021-00358-x#citeas.

4 'UK Could Cut Food Emissions by 17% by Sticking to a Healthy Diet' (2017), www.carbonbrief.org/uk-could-cut-food-emissions-17-per-cent-by-sticking-to-healthy-diet.

5 Stancu, V., Haugaard, P., and Lähteenmäki, L., 'Determinants of Consumer Food Waste Behaviour: Two Routes to Food Waste', *Appetite*, 96 (2016), pp. 7–17.

6 MIT Climate Portal, 'Fertilisers and Climate Change', climate.mit.edu/explainers/fertilizer-and-climate-change.

7 'EAT-Lancet Commission Summary Report', eatforum.org/eat-lancet-commission/eat-lancet-commission-summary-report/.

8 'The Occurrence of Selected Hydrocarbons in Food on Sale at Petrol Station Shops', Report, Brussels (I2000), www.concawe.eu/wp-content/uploads/2017/01/2002-00234-01-e.pdf.

9 'France to Eliminate Plastic Packaging from Fruits and Vegetables', Hunter College, New York City Food Policy Centre (2021), www.nycfoodpolicy.org/food-policy-snapshot-france-ban-plastic-packaging-fruits-and-vegetables/.

10 China CSA Network, Transformative Cities Programme, transformativecities.org/atlas/atlas-51/.

11 'Nearly Half Our Calories Come from Just 3 Crops. This Needs to Change', World Economic Forum (2018), www.weforum.org/agenda/2018/10/once-neglected-these-traditional-crops-are-our-new-rising-stars.

12 Holstein Cattle Fact Sheet, Holstein Association USA, www.holsteinusa.com/pdf/fact_sheet_cattle.pdf.

13 Poore, J. Nemeck, T, 'Reducing Food's Environmental Impacts Through Producers and Consumers', *Science* vol. 360, no. 6392 (2018).

14 Bedingfield, Will, 'Lab-grown Tuna Steaks Could Reel in Our Overfishing Problem', *Wired* (2021), www.wired.co.uk/article/blue-nalu-lab-grown-fish.

Chapter 10

1   Eco Living archive, Eco Age, eco-age.com/resources/jane-goodalls-best-quotes/.

2   'Heglar', Mary Annaïse, *Hot Take* Newsletter (30 Jan. 2022)

3   Polo, S., 'Nausicaä of the Valley of the Wind Reminds Us that Everything Changes, and Life Goes On', www.polygon.com/animation-cartoons/2020/5/25/21265521/nausicaa-of-the-valley-of-the-wind-studio-ghibli-movie-watch-meaning-manga.

4   'The Green Belt Movement, and the Story of Wangari Maathai', *Yes* magazine (2005), www.yesmagazine.org/issue/media/2005/03/26/the-green-belt-movement-the-story-of-wangari-maathai.

5   Winfrey, Oprah, Wangari Maathai obituary, *Time* magazine (14 Dec. 2011), content.time.com/time/specials/packages/article/0,28804,2101745_2102136_2102234,00.html

6   Foster Botanical Garden Fact Sheet, www.honolulu.gov/parks/hbg/honolulu-botanical-gardens/182-site-dpr-cat/568-foster-botanical-garden.html.

7   'Erin Brockovich: The Real Story of the Town Three Decades Later', *ABC News* (2021), abcnews.go.com/US/erin-brockovich-real-story-town-decades/story?id=78180219.

8   www.sustainable-markets.org/terra-carta/.

9   Strassburg, B. B. N., Iribarrem, A., Beyer, H. L., et al. 'Global Priority Areas for Ecosystem Restoration', *Nature,* vol. 586, p. 724–9 (2020), www.nature.com/articles/s41586-020-2784-9.

10   Rewild.org launch press release (22 May 2021), www.rewild.org/team/leonardo-dicaprio.

BIBLIOGRAPHY

The following bibliography is just a snapshot of some of the Earth-focused books and podcasts I have encountered (and two that I've written!) that I think you might also enjoy.

Barber, Aja, *Consumed: The Need for Collective Change: Colonialism, Climate Change & Consumerism*, Octopus (2021)

Bell, Alice, *Our Biggest Experiment: A History of Climate Change*, Bloomsbury Sigma (2021)

Carson, Rachel, *Silent Spring*, Penguin Modern Classics (1962)

Earle, Sylvia A., *National Geographic Ocean: A Global Odyssey*, National Geographic (2021)

Goodall, Jane, *The Book of Hope: A Survival Guide for an Endangered Planet*, Penguin Books (2021)

Holthaus, Eric, *The Future Earth: A Radical Vision for What's Possible in the Age of Warming*, HarperCollins (2020)

Johnson, Ayana Elizabeth, and Wilkinson, Katharine K., *All We Can Save: Truth, Courage, and Solutions for the Climate Crisis*, Penguin Random House (2021)

Klein, Naomi, *This Changes Everything: Capitalism vs the Climate*, Simon & Schuster (2014)

Lang, Tim, *Feeding Britain: Our Food Problems and How to Fix Them*, Penguin (2021)

Lee, Sam, *The Nightingale*, Penguin (2021)

Maathai, Wangari, *The Green Belt Movement: Sharing the Approach and the Experience*, Lantern Books (2003)

McAnulty, Dara, *Diary of a Young Naturalist*, Ebury Press (2021)

Minney, Safia, *Slave to Fashion*, New Internationalist (2017)

Monbiot, George, *Feral: Rewilding the Land, Sea and Human Life*, Allen Lane (2013)

Nelles, D., and Serrer, C., *This is Climate Change: A Visual Guide to the Facts*, Experiment (2021)

Penniman, Leah, *Farming While Black: Soul Fire Farm's Practical Guide to Liberation on the Land*, Chelsea Green Publishing (2018)

Pollan, Michael, *In Defense of Food: An Eater's Manifesto*, Penguin Press (2008)

Porritt, Jonathon, *Hope in Hell: A Decade to Confront the Climate Emergency*, Simon & Schuster (2020)

Sheldrake, Merlin, *Entangled Life: How Fungi Make Our Worlds, Change Our Minds & Shape Our Futures*, Random House (2020)

Siegle, Lucy, *To Die For: Is Fashion Wearing Out the World?*, Fourth Estate (2011)

Siegle, Lucy, *Turning the Tide on Plastic: How Humanity (And You) Can Make Our Globe Clean Again*, Trapeze (2018)

Simard, Suzanne, *Finding the Mother Tree: Discovering the Wisdom of the Forest*, Allen Lane (2021)

Thunberg, Greta, *No One is Too Small to Make a Difference*, Penguin (2019)

Waldron, Sangeeta, *Corporate Social Responsibility Is Not Public Relations*, LID Publishing (2019)

Washington, Harriet A., *A Terrible Thing to Waste: Environmental Racism and Its Assault on the American Mind*, Hachette (2019)

Williams, Tracey, *Adrift: The Curious Tale of the Lego Lost at Sea*, Unicorn (2022)

INDEX